MW00876830

Soldier's Manual of Common Tasks: Warrior Skills Level 1

August 2015

United States Government
US Army

Contents

Preface v

Chapter 1 Introduction to the SMCT System 1-1

Chapter 2 Training Guide 2-1

Chapter 3 Warrior Skills Level 1 Tasks 3-1

Appendix A Battle Drills A-1

Appendix B Proponent School or Agency Codes B-1

Glossary Glossary-1

References References-1

Subject Area 1: Shoot/Maintain, Employ, and Engage Targets with
Individually Assigned Weapon System 3-1
071-COM-0032 Maintain an M16- series Rifle/M4 series Rifle
Carbine 3-1
071-COM-0029 Perform a Function Check on an M16- Series
Rifle/ M4 Series Carbine 3-9
071-COM-0028 Load an M16- Series Rifle/M4 Series Carbine 3-11
071-COM-0027 Unload an M16 Series Rifle/M4 Series
Carbine 3-14
071-COM-0030 Engage Targets with an M16-Series Rifle/M4
Series Carbine 3-16
071-COM-0033 Correct Malfunctions of an M16- Series Rifle/M4
Series Carbine 3-18
071-COM-0031 Zero an M16- Series Rifle/M4 Series Carbine 3-20
071-COM-4401 Perform Safety Checks on Hand Grenades 3-25

i

071-COM-4407 Employ Hand Grenades...............................3-30
Subject Area 2: Move 3-35
071-COM-0541 Perform Exterior Movement Techniques during an Urban Operation...3-35
071-COM-0503 Move Over, Through, or Around Obstacles (Except Minefields) 3-41
071-COM-1000 Identify Topographic Symbols on a Military Map 3-44
071-COM-1001 Identify Terrain Features on a Map.................3-49
071-COM-1008 Measure Distance on a Map 3-57
071-COM-1002 Determine the Grid Coordinates of a Point on a Military Map 3-61
071-COM-1005 Determine a Location on the Ground by Terrain Association..3-69
071-COM-1012 Orient a Map to the Ground by Map-Terrain Association..3-71
071-COM-1011 Orient a Map Using a Lensatic Compass 3-73
071-COM-1003 Determine a Magnetic Azimuth Using a Lensatic Compass...3-79
071-COM-1006 Navigate from One Point on the Ground to another Point while Dismounted...3-84
071-COM-0501 Move as a Member of a Team.......................3-87
071-COM-0502 Move Under Direct Fire 3-90
071 COM-0510 React to Indirect Fire while Dismounted.........3-95
071-COM-0513 Select Hasty Fighting Positions....................3-97
Subject Area 3: Communicate 3-99
113-COM-2070 Operate SINCGARS Single-Channel (SC) 3-99
113-COM-1022 Perform Voice Communications 3-101
171-COM-4079 Send a Situation Report (SITREP) 3-103
171-COM-4080 Send a Spot Report (SPOTREP) 3-105
071-COM-0608 Use Visual Signaling Techniques..................3-108
Subject Area 4: Survive 3-140
031-COM-1036 Maintain Your Assigned Protective Mask.....3-140
031-COM-1035 Protect Yourself from Chemical and Biological (CB) Contamination Using Your Assigned Protective Mask ..3-142
031-COM-1019 React to Chemical or Biological (CB) Hazard/Attack 3-150

031-COM-1040 Protect Yourself from CBRN Injury/Contamination with the JSLIST Chemical-Protective Ensemble **3-155**

031-COM-1013 Decontaminate Yourself and Individual Equipment Using Chemical Decontaminating Kits **3-159**

031-COM-1037 Detect Chemical Agents Using M8 or M9 Detector Paper...**3-167**

031-COM-1021 Mark CBRN-Contaminated Areas.................**3-170**

031-COM-1018 React to Nuclear Hazard/Attack....................**3-174**

031-COM-1042 Protect Yourself from CBRN injury/contamination when changing MOPP using JSLIST Chemical Protective Ensemble...**3-177**

081-COM-1044 Perform First Aid for Nerve Agent Injury...**3-184**

081-COM-1001 Evaluate a Casualty (Tactical Combat Casualty Care) **3-191**

081-COM-1003 Perform First Aid to Clear an Object Stuck in the Throat of a Conscious Casualty **3-196**

081-COM-1005 Perform First Aid to Prevent or Control Shock **3-198**

081-COM-1023 Open An Airway...**3-201**

081-COM-1032 Perform First Aid for Bleeding of an Extremity..**3-205**

081-COM-1046 Transport a Casualty.....................................**3-208**

081-COM-1007 Perform First Aid for Burns **3-216**

081-COM-1026 Perform First Aid for an Open Chest Wound **3-220**

081-COM-0101 Request Medical Evacuation..........................**3-223**

052-COM-1270 React to Possible Improvised Explosive Device (IED) Attack (Located at https://www.us.army.mil/suite/doc/23838478) **3-229**

052-COM-1271 Identify Visual Indicators of an Implosive Device (Located at https://www.us.army.mil/suite/doc/23838510) **3-229**

071-COM-0804 Perform Surveillance without the Aid of Electronic Device **3-229**

301-COM-1050 Report Information of Potential Intelligence Value **3-234**

071-COM-0815 Practice Noise, Light, and Litter Discipline ..**3-238**

071-COM-0801 Challenge Persons Entering Your Area.........**3-240**

071-COM-0512 Perform Hand-to-Hand Combat **3-243**
071-COM-4408 Construct Individual Fighting Position..........**3-258**
052-COM-1361 Camouflage Yourself and Your Individual
Equipment...**3-272**
071-COM-0011 Employ Progressive Levels of Individual
Force.. **3-279**
181-COM-1001 Conduct Operations According to the
Law of War **3-281**
191-COM-0008 Search an Individual in a Tactical
Environment.. **3-284**
159-COM-2026 Identify Combatant and Non-Combatant Personnel
& Hybrid Threats...**3-291**

Preface

This manual is one of a series of soldier training publications (STPs) that support individual training. Commanders, trainers, and Soldiers will use this manual and STP 21-24-SMCT to plan, conduct, sustain, and evaluate individual training of warrior tasks and battle drills in units.

This manual includes the Army Warrior Training plan for warrior skills level (SL) 1 and task summaries for SL 1 critical common tasks that support unit wartime missions. This manual is the only authorized source for these common tasks. Task summaries in this manual supersede any common tasks appearing in military occupational specialty (MOS)-specific Soldier manuals.

Training support information, such as reference materials, is also included. Trainers and first-line supervisors will ensure that SL 1 Soldiers have access to this publication in their work areas, unit learning centers, and unit libraries.

This manual applies to the Active Army, the Army National Guard/Army National Guard of the United States, and the U.S. Army Reserve unless otherwise stated.

The proponent of this publication is the United States Army Training and Doctrine Command (TRADOC), with the United States Army Training Support Center (ATSC) designated as the principle publishing, printing, and distribution agency. Proponents for the specific tasks are the Army schools and agencies as identified by the school code, listed in appendix A. This code consists of the first three digits of the task identification number.

Record any comments or questions regarding the task summaries contained in this manual on a DA Form 2028 (*Recommended Changes to Publications and Blank Forms*) and send it to the respective task proponent, James Rose CIMT, james.a.rose20.civ@mail.mil, with information copies forwarded to—

> • Commander, U.S. Army Training and Doctrine Command,
> ATTN: ATCG-MT,
> Fort Eustis, VA 23604-5701.

> • Commander, U.S. Army Training Support Center,
> ATTN: ATIC-APR,
> Fort Eustis, VA 23604-5166.

The Soldier's Creed

I am an American Soldier.

I am a Warrior and a member of a team.
I serve the people of the United States and live the Army Values.

I will always place the mission first.

I will never accept defeat.

I will never quit.

I will never leave a fallen comrade.

I am disciplined, physically and mentally tough, trained, and
proficient in my warrior tasks and drills.
I always maintain my arms, my equipment, and myself.

I am an expert and I am a professional.

I stand ready to deploy, engage, and destroy
the enemies of the United States of America
in close combat.

I am a guardian of freedom and
the American way of life.

I am an American Soldier.

CHAPTER 1

Introduction to the SMCT System

1-1. GENERAL

The Army's basic mission is to train and prepare Soldiers, leaders, and units to fight and win in combat. As explained in the Army's capstone training doctrine (ADP 7-0), units do not have the time or the resources to achieve and sustain proficiency with every possible training task. Therefore, commanders must identify the tasks that are the units' critical wartime tasks. These tasks then become the unit's Mission Essential Task List (METL). Commanders use the METL to develop their unit-training plan. Noncommissioned officers (NCOs) plan the individual training that Soldiers need to become warriors and to accomplish the METL. The STPs, also known as Soldier's manuals (SMs), provide the critical individual tasks for each military occupational specialty (MOS) that support all of the unit's missions. The NCO leadership uses the tasks in the SMs to train the Soldiers and measure the Soldiers' proficiency with these unit-critical tasks. The manuals provide task performance and evaluation criteria and are the basis for individual training and evaluation in the unit and for task-based evaluation during resident training.

The Army identified warrior tasks and battle drills (WTBD) that enhance a Soldier's readiness to fight on the battlefield. Warrior tasks are a collection of individual Soldier skills known to be critical to Soldier survival. Examples include weapons training, tactical communications, urban operations, and first aid. Battle drills are group skills designed to teach a unit to react and survive in common combat situations. Examples included react to ambush, react to chemical attack, and evacuate injured personnel from a vehicle. WTBD increases the relevance of training to current combat requirements and enhance the rigor in training. The driving force behind the change comes from lessons learned. Standards remain constant but commanders must be aware that the enemy adapts at once and Soldier training will change sooner because of current operational environments.

Note: If a task identified in the SMCT is not current refer to "DTMS", or the Central Army Registry (CAR) https://atiam.train.army.mil/catalog/catalog/search.html, all tasks are reviewed annually and may change before the SMCT is updated.

1-2. PURPOSE

This Soldier's Manual of Common Tasks (SMCT), Warrior Skill Level (SL) 1, contains the individual tasks that are essential to the Army's ability to win on the modern battlefield. In an operational environment, regardless of job or individual MOS, each Soldier risks exposure to hostile actions. This manual contains the warrior skills that Soldiers must be able to perform to fight, survive, and win in combat.

This SMCT gives the commander, NCO trainer, first-line supervisor, and individual Soldiers the information necessary to support integration and sustainment training in their units. This information allows trainers to plan, prepare, train, evaluate, and monitor individual training of warrior tasks. Using the appropriate mission-training plan (MTP), military occupational specialty (MOS)-specific Soldier's training publication (STP), and this manual helps provide the foundation for an effective
unit-training plan.

1-3. COMMANDER'S RESPONSIBILITIES

The commander at each level develops a unit METL in consultation with the command sergeant major and subordinate commanders. Using the training planning process described in ADP 7-0, the commander develops the METL and then determines the level of training needed to attain and maintain proficiency. WTBD in Chapter 3 supports an Army at war and becomes the key element in Army Warrior Training (AWT). Commanders use the unit METL and AWT to determine the necessary training for the unit and develop a strategy to accomplish the required training throughout the fiscal year (FY). The commander also gives the NCO leadership the guidance they need to carry out this strategy. Each commander must design a unit training plan that prepares the unit for the full spectrum of operations. Soldiers must develop and sustain proficiency in the critical tasks for their MOS and skill level. The commander's unit training program should provide individual training for all Soldiers assigned to the unit and evaluate Soldier proficiency by routine. The leader's assessment and the AWT are two tools that give the NCO leadership and commander information about the status of training for individuals and for the unit, which should be integrated with collective training such as the MTPs, crew drills, and battle drills.

Chapter 2 provides information about where tasks are trained to standard and how often tasks are trained to maintain proficiency.

Based on the commander's guidance, individual training in the unit is the responsibility of the NCO trainers. The commander must give the NCO trainer the priorities, resources, and directions needed to carry out training. He or she must also assess the training results of the MTP and other training events, and adjust the unit training plan as a result. To develop a training program, use the following seven-step approach:

> **Step 1.** Set the objectives for training.
>
> **Step 2.** Plan the resources (personnel, time, funds, facilities, devices, and training aids).
>
> **Step 3.** Train the trainers.
>
> **Step 4.** Provide the resources.
>
> **Step 5.** Manage risks, environmental and safety concerns.
>
> **Step 6.** Conduct the training.
>
> **Step 7.** Evaluate the results.

1-4. TRAINER'S RESPONSIBILITIES

Trainers must use the following steps to plan and evaluate training:

a. *Identify individual training requirements.* The NCO determines which tasks Soldiers need to train based on the commander's training strategy. The unit's training plan, METL, MTP, and the AWT plan (Chapter 2) are sources for helping the trainer define the individual training needed.

b. *Plan the training.* Plan individual training based on the unit's training plan. Be prepared to take advantage of opportunities to conduct individual training ("hip pocket" training).

c. *Gather the training references and materials.* The task summaries list references that can assist the trainer in preparing for the training of that task. The Reimer Digital Library provides current training materials.

d. *Manage risks and environmental and safety concerns.* Assess the risks involved with training a specific task regarding the conditions current at the time of training and, if necessary, implement controls to reduce the risk level. Ensure that training preparation takes into account those cautions, warnings, and dangers associated with each task as well as environmental and safety concerns (ATP 5-19).

e. *Train each Soldier.* Demonstrate to the Soldier how to do the task with standard proficiency and explain (step by step) how to do the task. Give each Soldier the opportunity to practice the task step by step.

f. *Check each Soldier.* Evaluate how well each Soldier performs the tasks in this manual. Conduct these evaluations during individual training sessions or while evaluating individual proficiencies when conducting unit collective tasks. This manual provides a training and evaluation guide for each task to enhance the NCO's ability to conduct year-round, hands-on evaluations of tasks critical to the unit's mission. Use the information in the AWT plan (chapter 2) as a guide to determine how often to train Soldiers using each task to maintain proficiency.

g. *Record the results.* Use the leader book referred to in ADP 7-0 to record task performance. This gives the leader total flexibility with the methods of recording training tasks. The trainer may use DA Form 5164-R (*Hands-on Evaluation*) and DA Form 5165-R (*Field Expedient Squad Book*) as part of the leader book. These forms are optional and reproducible anywhere.

h. *Retrain and evaluate.* Work with each Soldier until he/she performs the task to standard. Well-planned, integrated training increases the professional competence of each Soldier and contributes to the development of an efficient unit. The NCO or first-line supervisor is a vital link to the conduct of training.

1-5. SOLDIER'S RESPONSIBILITIES

Each Soldier must be able to perform the individual tasks that the first-line supervisor has identified based on the unit's METL. The Soldier must perform the task to the standard listed in this SMCT. If a Soldier has a question about how to do a task, or which tasks in this manual he or she must perform, it is the Soldier's responsibility to go to the first-line supervisor for clarification. The first-line supervisor knows how to perform each task or can direct the Soldier to the appropriate training materials. In addition, each Soldier should—

a. Know the training steps for both the WTBD and the MOS-specific critical tasks for his or her skill level. A list of the critical tasks is found in chapter 2 of this manual and the STP for the specific MOS (MOS-specific tasks).

b. Check the Reimer Digital Library for new training materials to support self-development with maintaining earlier trained tasks or to learn new tasks.

1-6. TASK SUMMARIES

Task summaries document the performance requirements of a critical warrior task. They provide the Soldier and the trainer with the information necessary to evaluate critical tasks. The formats for the task summaries are—

a. *Task title.* The task title identifies the action to perform.

b. *Task number.* The task number is an l0-digit number that identifies each task. The first three digits of the number represent the proponent code for that task. (Appendix A provides a list of proponent codes.) Include the entire 10-digit task number, along with the task title, in any correspondence relating to the task.

c. *Conditions.* The task conditions identify all the equipment, tools, materials, references, job aids, and supporting personnel that the Soldier needs to perform the task. This section identifies any environmental conditions that can alter task performance such as visibility, temperature, or wind. This section also identifies any specific cues or events (for example, a chemical attack or identification of an unexploded ordnance hazard) that trigger task performance.

d. *Standards.* A task standard specifies the requirements for task performance by indicating how well, complete, or accurate a product must be produced, a process must be performed, or both. Standards are described in terms of accuracy, tolerances, completeness, format, and clarity, number of errors, quantity, sequence, or speed of performance.

e. *Training and evaluation guide.* This section has two parts. The first part, Performance Steps, lists the individual steps that the Soldier must complete to perform the task. The second part is the Performance Evaluation Guide. This provides guidance about how to evaluate a Soldier's performance of the task. It is composed of three subsections. The *Evaluation Preparation* subsection identifies special setup procedures and, if required, instructions for evaluating the task performance. Sometimes the conditions and standards must be modified so that the task can be evaluated in a situation that does not, without approximation, duplicate actual field performance. The *Performance Measures* subsection identifies the criteria for acceptable task performance. The Soldier is rated (GO/NO- GO) on how well he or she performs specific actions or produces specific products. As indicated in *Evaluation Guidance*, a Soldier must score a GO on all or specified performance measures to receive a GO on the task in order to be considered trained.

f. *References.* This section identifies references that provide more detailed and thorough explanations of task performance requirements than that given in the task summary description. This section identifies resources the Soldier can use to improve or maintain performance.

g. In addition, task summaries can include safety statements, environmental considerations, and notes. Safety statements (danger, warning, and caution) alert users to the possibility of immediate death, personal injury, or damage to equipment. Notes provide additional information to support task performance.

1-7. TRAINING TIPS FOR NCO LEADERS

a. Prepare yourself.

(1) Get training guidance from your chain of command about when to train, which Soldiers to train, availability of resources, and a training site.

(2) Get task, conditions and standards from the task summary in this manual. Ensure that you can do the task. Review the task summary and the references in the reference section. Practice doing the task or, if necessary, have someone train you how to perform the task.

b. Prepare the resources.

(1) Obtain the required resources as identified in the conditions statement for each task and/or modified in the training and evaluation guide.

(2) Gather the equipment and ensure that it is operational.

(3) Prepare a training outline consisting of informal notes about what you want to cover during your training session.

(4) Practice your training presentation.

(5) Coordinate for the use of training aids and devices.

(6) Prepare the training site using the conditions statement as modified in the training and evaluation guide.

c. Train the Soldiers.

(1) Tell the Soldier what task to do and how well it must be done. Refer to the task standards and the performance measures for the task, as appropriate.

(2) Caution Soldiers about safety, environment, and security considerations.

(3) Demonstrate how to do the task to the standard level. Have the Soldiers study the appropriate training materials.

(4) Provide any necessary training involving basic skills Soldiers must have before they can be proficient with the task.

(5) Have the Soldiers practice the task until they can perform it to standard levels.

(6) Provide critical information to those Soldiers who fail to perform at task standard levels, and have them continue to practice until they can perform at standard levels.

(7) Combine training involving the individual tasks contained in this manual with the collective tasks contained in the MTP. Ensure that the necessary safety equipment and clothing needed for proper performance of the job are on hand at the training site.

d. Record the results: First-line supervisors record the results and report information to the unit leadership.

1-8. TRAINING SUPPORT

Appendix A lists the task proponents and agency codes (first three digits of the task number) with addresses for submitting comments concerning specific tasks in this manual.

1-9. EVALUATING TASK PERFORMANCE

Trainers need to keep the following points in mind when preparing to evaluate their Soldiers:

a. Review the performance measures to become familiar with the criteria about which you will score the Soldier.

b. Ensure that all necessary equipment and clothing needed for proper performance of the job are on hand at the training site. Remember to include safety equipment.

c. Prepare the test site according to the conditions section of the task summary. Some tasks contain special evaluation preparation instructions. These instructions tell the trainer what modifications must be made concerning job conditions to evaluate the task. Reset the site to its original condition after evaluating each Soldier to ensure that the conditions are the same for each Soldier.

d. Advise each Soldier about any special guidance that appears in the evaluation preparation section of the task summary before evaluating.

e. Score each Soldier regarding the information in the performance measures and evaluation guidance. Record the date of training and task performance score (GO / NO GO) in the sections training records for each Soldier.

(1) When applicable, conduct an exercise after-action review to allow training participants to discover for themselves what happened, why it happened, and how it can be done better. Once all key points are discussed and linked to future training, the evaluator will make the appropriate notes for inclusion into the score.

(2) Score the Soldier GO if all performance measures pass. Score the Soldier NO GO if the Soldier fails any step. If the Soldier fails, Show the Soldier what they did wrong and allow the Soldiers to take the test again.

This page intentionally left blank.

CHAPTER 2
Training Guide

2-1. THE ARMY WARRIOR TRAINING PLAN

a. Army Warrior Training focuses on training Soldiers warrior tasks, battle drills, and tasks from a unit's METL. This chapter and chapter 3 provides information identifying individual tasks to train and assist in the trainer's planning, preparation, training assessment, and monitoring of individual training in units. It lists by general subject area, and skill level, the critical warrior tasks Soldiers must perform, the initial training location, and a suggested expertise of training.

b. The training location column uses brevity codes to indicate where the task is first taught to standard levels. If the task is taught in the unit, the word "UNIT" appears in this column. If the task is trained by a self-development media, "SD" appears in this column. If the task is taught in the training base, the brevity code (BCT, OSUT, and AIT) of the resident course appears. Brevity codes and resident courses are listed below.

Brevity Codes	
BCT	Basic Combat Training
OSUT	One Station Unit Training
AIT	Advanced Individual Training
UNIT	Trained in/by the Unit
SD	Self-Development Training

c. The sustainment-training column lists how often (frequency) Soldiers should train with the task to ensure they maintain their proficiency. This information is a guide for commanders to develop a comprehensive unit training plan. The commander, in conjunction with the unit trainers, is in the best position to determine which tasks, and how often Soldiers should train to maintain unit readiness. (See chapter 3 for a list of individual tasks that support the WTBD to be trained in each Army unit.)

Frequency Codes	
AN	Annually
SA	Semiannually
QT	Quarterly

Army Warrior Training Plan			
Task Number	**Title**	**Training Location**	**Sustainment Training Frequency**
Warrior Skill Level 1			
Subject Area 1: Shoot/Maintain, Employ, and Engage Targets with Individually Assigned Weapon System:			
071-COM-0032	Maintain an M16 Series Rifle/M4 Series Rifle Carbine	BCT/OSUT	AN
071-COM-0029	Perform a Function Check on an M16-Series Rifle/M4 Series Carbine	BCT/OSUT	QT
071-COM-0028	Load an M16- Series/M4 Series Carbine	BCT/OSUT	SA
071-COM-0027	Unload an M16- Series Rifle/M4Series Carbine	BCT/OSUT	SA
071-COM-0030	Engage Targets with an M16-Series Rifle/ M4 Series Carbine	BCT/OSUT	SA
071-COM-0033	Correct Malfunctions of an M16-Series Rifle /M4 Series Carbine	BCT/OSUT	QT
071-COM-0031	Zero an M16-Series Rifle/M4 Series Carbine	BCT/OSUT	SA
Subject Area 2: Shoot/Employ Hand Grenades:			
071-COM-4401	Perform Safety Checks on Hand Grenades	BCT/OSUT	AN
071-COM-4407	Employ Hand Grenades	BCT/OSUT	AN
Subject Area 3: Move/ Perform Individual Movement Techniques:			
071-COM-0541	Perform Exterior Movement Techniques	BCT/OSUT	AN
071-COM-0503	Move Over, Through, or Around Obstacles (Except Minefields)	BCT/OSUT	SA
Subject Area 4: Move/ Navigate From One Point To Another:			
071-COM-1000	Identify Topographic Symbols on a Military Map	BCT/OSUT	AN
071-COM-1001	Identify Terrain Features on a Map	BCT/OSUT	AN
071-COM-1008	Measure distance on a Map	BCT/OSUT	AN

Army Warrior Training Plan			
Task Number	**Title**	**Training Location**	**Sustainment Training Frequency**
071-COM-1002	Determine the Grid Coordinates of a Point on a Military Map	BCT/OSUT	AN
071-COM-1005	Determine a Location on the Ground by Terrain Association	Unit	AN
071-COM-1012	Orient a Map to the Ground by Map-Terrain Association	BCT/OSUT	AN
071-COM-1011	Orient a Map Using a Lensatic Compass	BCT/OSUT	AN
071-COM-1003	Determine a Magnetic Azimuth Using a Lensatic Compass	BCT/OSUT	AN
071-COM-1006	Navigate from One Point on the Ground to Another Point While Dismounted	BCT/OSUT	SA
Subject Area 5: Move/ Move as a member of a Team:			
071-COM-0501	Move as a Member of a Team	BCT/OSUT	SA
071-COM-0502	Move Under Direct Fire	BCT/OSUT	SA
071-COM-0510	React to Indirect Fire While Dismounted	BCT/OSUT	SA
071-COM-0513	Select Hasty Fighting Positions	BCT/OSUT	SA
Subject Area 6: Communicate/ Perform Voice Communications (SITREP/SPOTREP/9-Line MEDEVAC):			
113-COM-2070	Operate SINCGARS Single-Channel (SC)	BCT/OSUT	SA
113-COM-1022	Perform Voice Communications	BCT/OSUT	AN
171-COM-4079	Send a Situation Report (SITREP)	BCT/OSUT	AN
171-COM-4080	Send a Spot Report (SPOTREP)	BCT/OSUT	AN
Subject Area 7: Communicate/ Visual Signaling Techniques:			
071-COM-0608	Use Visual Signaling Techniques	BCT/OSUT	SA

Army Warrior Training Plan			
Task Number	*Title*	*Training Location*	*Sustainment Training Frequency*
Subject Area 8: Survive/ React to Chemical, Biological, Radiological, and Nuclear (CBRN) Attack/Hazard:			
031-COM-1036	Maintain Your Assigned Protective Mask	BCT/OSUT	AN
031-COM-1035	Protect Yourself from Chemical and Biological (CB) Contamination Using Your Assigned Protective Mask	BCT/OSUT	AN
031-COM-1019	React to Chemical or Biological (CB) Hazard/Attack	BCT/OSUT	AN
031-COM-1040	Protect Yourself from CBRN Injury/Contamination with the JSLIST Chemical-Protective Ensemble	BCT/OSUT	SA
031-COM-1013	Decontaminate Yourself and Individual Equipment Using Chemical Decontaminating Kits	BCT/OSUT	AN
031-COM-1037	Detect Chemical Agents Using M8 or M9 Detector paper	BCT/OSUT	AN
031-COM-1021	Mark CBRN-Contaminated Areas	Unit	AN
031-COM-1018	React to Nuclear Hazard / Attack	BCT/OSUT	AN
031-COM-1042	Protect Yourself from CBRN injury/contamination when Changing MOPP using JSLIST	Unit	AN
081-COM-1044	Perform First Aid for Nerve Agent Injury	BCT/OSUT	AN
Subject Area 9: Survive/ Perform Immediate Lifesaving Measures:			
081-COM-1001	Evaluate a Casualty (Tactical Combat Casualty Care)	BCT/OSUT	AN

Army Warrior Training Plan			
Task Number	**Title**	**Training Location**	**Sustainment Training Frequency**
081-COM-1003	Perform First Aid to Clear an Object Stuck in the Throat of a Conscious Casualty	BCT/OSUT	AN
081-COM-1005	Perform First Aid to Prevent or Control Shock	BCT/OSUT	AN
081-COM-1023	Perform First Aid to Restore Breathing and/or Pulse	BCT/OSUT	AN
081-COM-1032	Perform First Aid for a Bleeding and/or Severed Extremity	BCT/OSUT	AN
081-COM-1046	Transport a Casualty	BCT/OSUT	AN
081-COM-1007	Perform First Aid for Burns	BCT/OSUT	AN
081-COM-1026	Perform First Aid for an Open Chest Wound	BCT/OSUT	AN
081-COM-0101	Request Medical Evacuation	BCT/OSUT	AN
Subject Area 10: Survive/ Perform Counter IED:			
052-COM-1270	React to Improvised Explosive Device (IED)	BCT/OSUT	AN
052-COM-1271	Identify Visual Indicators of an Improvised Device (IED)	BCT/OSUT	AN
Subject Area 11: Survive/ Maintain Situational Awareness:			
071-COM-0804	Perform Surveillance without the Aid of Electronic Device	Unit	SA
301-COM-1050	Report Information of Potential Intelligence Value	BCT/OSUT	SA
071-COM-0815	Practice , Noise, Light, and Litter Discipline	BCT/OSUT	SA
071-COM-0801	Challenge Persons Entering your Area	BCT/OSUT	AN
Subject Area 12: Survive/ Perform Combatives:			
071-COM-0512	React to Hand-to-Hand Combat	BCT/OSUT	SA
Subject Area 13: Survive/Construct an Individual Fighting Position			
071-COM-4408	Construct an Individual Fighting Position	BCT/OSUT	AN

Army Warrior Training Plan			
Task Number	**Title**	**Training Location**	**Sustainment Training Frequency**
051-COM-1361	Camouflage Yourself and Individual Equipment	BCT/OSUT	AN
171-COM-0011	Employ Progressive Levels of Individual Force	BCT/OSUT	AN
181-COM-1001	Conduct operations According to the Law of War	BCT/OSUT	AN
191-COM-0008	Search an Individual in a Tactical Environment	BCT/OSUT	AN
159-COM-2026	Identify Combatant and Non-Combatant Personnel & Hybrid Threats	BCT/OSUT	AN
Subject Area 14: (Battle Drills) React to Contact:			
071-COM-0513	Select Hasty Fighting Positions (Repeat)	BCT/OSUT	SA
071-COM-0030	Engage Targets with an M16-Series Rifle/ M4 Series Carbine (Repeat)	BCT/OSUT	SA
071-COM-0608	Use Visual Signaling Techniques (Repeat)	BCT/OSUT	AN
071-COM-0502	Move under Direct Fire (Repeat)	BCT/OSUT	SA
071-COM-0510	React to Indirect Fire While Dismounted(Repeat)	BCT/OSUT	SA
113-COM-1022	Perform Voice Communications (Repeat)	BCT/OSUT	AN
071-COM-0501	Move as a member of a Team (Repeat)	BCT/OSUT	SA
071-COM-4407	Employ Hand Grenades (Repeat)	BCT/OSUT	AN
052-COM-1271	Identify Visual Indicators of an Improvised Explosive Device (Repeat)	BCT/OSUT	AN
051-COM-1270	React to Possible Improvised Explosive Device (Repeat)	BCT/OSUT	AN
191-COM-5148	Search an Individual in a Tactical Environment	BCT/OSUT	AN
Subject Area 15: (Battle Drills) Establish Security at the HALT:			
071-COM-0513	Select Hasty Fighting Positions (Repeat)	BCT/OSUT	SA

Army Warrior Training Plan			
Task Number	Title	Training Location	Sustainment Training Frequency
113-COM-1022	Perform Voice Communications (Repeat)	BCT/OSUT	AN
071-COM-0801	Challenge Persons Entering Your Area	BCT/OSUT	AN
071-COM-1004	Perform Duty as a Guard	BCT/OSUT	AN
071-COM-0815	Practice Noise, Light, and Litter Discipline (Repeat)	BCT/OSUT	AN
191-COM-5148	Search an Individual in a Tactical Environment	Unit	AN
071-COM-0608	Use Visual Signaling Techniques (Repeat)	BCT/OSUT	AN
071-COM-4080	Send a SPOT Report	BCT/OSUT	AN
071-COM-4079	Send a SITREP Report	BCT/OSUT	AN
051-COM-1361	Camouflage Yourself and Individual Equipment	BCT/OSUT	AN
071-COM-4408	Construct an Individual Fighting Position	BCT/OSUT	AN
Subject Area 16: (Battle Drills) Perform Tactical Combat Casualty Care:			
081-COM-0101	Request Medical Evacuation (Repeat)	BCT/OSUT	SA
081-COM-1001	Evaluate a Casualty (Repeat)	BCT/OSUT	AN
081-COM-1003	Perform First Aid to Clear an Object Stuck in the Throat of a Conscious Casualty (Repeat)	BCT/OSUT	AN
081-COM-1005	Perform First Aid to Prevent or Control Shock (Repeat)	BCT/OSUT	AN
081-COM-1023	Open an Airway	BCT/OSUT	AN
081-COM-1032	Perform First Aid for Bleeding of an Extremity (Repeat)	BCT/OSUT	AN
081-COM-1046	Transport a Casualty (Repeat)	BCT/OSUT	AN
113-COM-1022	Perform Voice Communications (Repeat)	BCT/OSUT	AN
081-COM-1054	Evacuate casualties	BCT/OSUT	AN
191-COM-5148	Search an Individual in a tactical Environment	BCT/OSUT	AN

Army Warrior Training Plan			
Task Number	*Title*	*Training Location*	*Sustainment Training Frequency*
Subject Area 17: React to Ambush (near):			
052-COM-1271	Identify Visual Indicators of an IED (Repeat)	BCT/OSUT	AN
052-COM-3261	React to an IED Attack (Repeat)	BCT/OSUT	AN
071-COM-0006	React-to-Hand-to-Hand-Combat (Repeat)	BCT/OSUT	AN
071-COM-0030	Engage Targets with M4/M16 Rifle (Repeat)	BCT/OSUT	SA
071-COM-4407	Employ Hand Grenades (Repeat)	BCT/OSUT	AN
071-COM-0501	Move as a member of a team (Repeat)	BCT/OSUT	AN
071-COM-0502	Move under direct fire (Repeat)	BCT/OSUT	AN
071-COM-0513	Select Hasty fighting positions (Repeat)	BCT/OSUT	AN
071-COM-0608	Use visual Signaling Techniques (Repeat)	BCT/OSUT	AN
113-COM-1022	Perform voice communication (Repeat)	BCT/OSUT	AN
Subject Area 18: React to Ambush (far):			
052-COM-1270	React to Possible Improvised Explosive Device (Repeat)	BCT/OSUT	AN
071-COM-0501	Move as a Member of a Team	BCT/OSUT	SA
071-COM-0513	Select Hasty Fighting Positions (Repeat)	BCT/OSUT	SA
113-COM-1022	Perform Voice Communications (Repeat)	BCT/OSUT	AN
071-COM-0608	Use Visual Signaling Techniques (Repeat)	BCT/OSUT	AN
071-COM-0030	Engage Targets with M4/M16 Rifle (Repeat)	BCT/OSUT	SA

CHAPTER 3

Warrior Skills Level 1 Tasks

Subject Area 1: Shoot/Maintain, Employ, and Engage Targets with Individually Assigned Weapon System

071-COM-0032

Maintain an M16 Series Rifle/M4 Series Rifle Carbine

WARNING
Do not squeeze the trigger until the weapon has been cleared. Inspect the chamber to ensure that it is empty and no ammunition is in position to be chambered. Failure to do so may lead to death or serius injury.

Conditions:
You have just returned from a mission with your loaded M16 series rifle or M4 series carbine and have been directed to conduct maintenance on your weapon. You have a small-arms accessory case. Some iterations of this task should be performed in MOPP 4.

Standards: Clear, disassemble, clean, inspect, lubricate, assemble, and perform a function check on the M16/M4. Maintain the magazine and ammunition.

Special Condition: This task is being superseded by 071-COM-0033, REFER TO DTMS OR CAR (https://atiam.train.army.mil/catalog/catalog/search.html) for the new task

Special Standards: None

Special Equipment:

Cue: None

DANGER
Do not squeeze the trigger until the weapon has been cleared. Inspect the chamber to ensure that it is empty and no ammunition is in position to be chambered. Failure to do so may lead to death or serious injury.

Note: None

WARNING

Weapon must be cleared to be considered safe.

---

1. Clear the weapon.
   a. Point weapon in safe direction.
   b. Attempt to place the selector lever on SAFE.
Note: If weapon is not cocked, lever can't be pointed toward safe.
   c. Remove the magazine from the weapon, if present.
   d. Lock the bolt open.
      (1) Pull the charging handle rearward.
      (2) Press the bottom of the bolt catch.
      (3) Move the bolt forward until it engages the bolt catch.
      (4) Return the charging handle to the forward position.
      (5) Ensure the receiver and chamber are free of ammo.
   e. Place the selector lever on safe.
   f. Press the upper portion of the bolt catch to allow the bolt to go forward.
   g. Place the selector lever from SAFE to SEMI.
   h. Squeeze trigger.
   i. Pull the charging handle fully rearward and release it, allowing the bolt to return to the full forward position.
   j. Place the selector lever on SAFE.
2. Disassemble the weapon.
   a. Remove the sling.

CAUTION

Do not use a screwdriver or any other tool when removing the handguards. Doing so may damage the handguards, slip ring, or both.

Do not bend or dent the gas tube while removing handguard.

---

   b. Remove the handguards only if you can see dirt or corrosion through the vent holes.
Note: Hand guards on the M16A2 are interchangeable because they are identical. On the M16A4 the hand guards can be replaced by the M5 adapter rails. On the M4 carbine series, the hand guards can be replaced by the M4 adapter rails. The M4 and M5 adapter rails are marked with a T for top and B for bottom. The operator is only authorized to remove the lower adapter rail and rail covers for cleaning, lubrication, or attaching accessories.
      (1) Place the weapon on the buttstock.
      (2) Press down on the slip ring with both hands.
      (3) Pull the handguards free.
   c. Push the take down pin as far as it will go.
   d. Pivot the upper receiver from the lower receiver.
   e. Push the receiver pivot pin in as far as it will go.
   f. Separate the upper and lower receivers.
   g. Remove carrying handle, if applicable.
      (1) Loosen the screws on the left side of the clamping bar.

(2) Lift the handle off once the clamping bar is loose.

h. Pull back the charging handle.

i. Remove the bolt carrier and bolt.

j. Remove the charging handle.

k. Disassemble the bolt carrier.

(1) Remove the firing pin retaining pin.

Note: Do not spread open or close split end of pin.

(2) Push in bolt assembly to locked position.

CAUTION
Do not drop or hit the firing pin. Damage to the pin may cause the weapon to malfunction.

(3) Drop firing pin out of rear of bolt carrier.

(4) Remove the bolt cam pin by turning it one-quarter of a turn and lifting it out.

(5) Remove bolt assembly from carrier.

(6) Press the rear of the extractor pin to check spring function.

Note: Any weak springs should be reported to the unit armor for replacement.

(7) Remove the extractor pin by pushing it out with the firing pin.

(8) Lift out the extractor and spring, taking care that the spring does not separate from the extractor.

l. Remove buffer and buffer spring from buttstock.

(1) Press in buffer depress retainer and release buffer.

(2) Remove buffer and action spring.

m. Remove the buttstock. (M4 series only)

(1) Extend the buttstock assembly to full open.

(2) Separate the buttstock assembly from the lower receiver extension.

(a) Grasp the lock lever in the area of the retaining nut.

(b) Pull downward.

(c) Slide the buttstock to the rear.

3. Clean the weapon.

Note: CLP is used to identify when lubricant is needed, however it can be replaced with LSA (weapons lubricant oil, semifluid), or LAW (lubricating oil, arctic weather) as applicable.

Do not mix lubricants on the same weapon. The weapon must be thoroughly cleaned using dry cleaning solvent (SD) when changing from one lubricant to another.

CAUTION
Do not mix parts of one weapon with other weapons. Parts are not interchangeable.

a. Clean the bore.

Note: The bore of your weapon has lands and grooves called rifling. Rifling makes the bullet spin very fast as it moves down the bore and down range. Because it twists so quickly, it is difficult to push a new, stiff bore brush through the bore. You will find it easier to pull your bore brush through the bore. Also, because the brush will clean better if the bristles follow the grooves (called tracking), you want the bore brush to be allowed to turn as you pull it through.

(1) Attach three cleaning rod sections together.

(2) Swab out the bore with a patch moistened with CLP or RBC.

(3) Attach the bore brush.

Note: When using bore brush, don't reverse direction while in bore.

(4) Point muzzle down.

(5) Hold the upper receiver in one hand while inserting the end of the rod without the brush into the chamber.

(6) Let the rod fall straight through the bore.

Note: About 2 to 3 inches will be sticking out of the muzzle at this point.

(7) Attach the handle section of the cleaning rod to the end of the rod sticking out of the muzzle.

(8) Pull the brush through the bore and out of the muzzle.

(9) Take off the handle section.

(10) Run the brush through the bore again by repeating the process.

(11) Replace the bore brush with the rod tip.

(12) Attach a patch with CLP to the rod tip.

(13) Pull the patch through the bore.

b. Upper receiver group.

(1) Connect chamber brush to cleaning rod handle.

(2) Dip the chamber brush in CLP and insert in chamber and locking lugs.

(3) Push and twist to clean.

(4) Use a worn out bore brush to clean outside of gas tube.

Note: Gas tubes will discolor from heat. Do not attempt to remove discoloration.

(5) Clean the entire upper receiver by wiping it down.

c. Bolt carrier group.

(1) Clean carbon and oil from firing pin.

(2) Clean bolt carrier key with worn brush.

(3) Clean firing pin recess with pipe cleaner.

(4) Clean firing pin hole with pipe cleaner.

(5) Clean behind bolt rings and lip of extractor.

(6) Clean carbon deposits and dirt from locking lugs.

CAUTION

Do not use wire brush or any other type of abrasive material to clean aluminum surfaces. Damage to equipment may occur.

---

d. Lower receiver group.

(1) Wipe dirt from trigger with a patch.

(2) Use a patch dampened with CLP to clean powder fouling, corrosion, and dirt from outside parts of lower receiver and extension assembly.

(3) Use pipe cleaner to clean buttstock drain hole.

(4) Clean buffer assembly, spring, and inside with patch dampened with CLP.

(5) Wipe dry.

e. Clean the ejector.

(1) Place a few drops of CLP on the ejector.

(2) Press the ejector in using a spent round casing or dummy round.

(3) Hook casing under extractor and rock back and forth against ejector.

(4) Repeat this process a few times adding lubricant until the action of the ejector is smooth and strong.

(5) Dry off excess CLP when process is completed.

4. Inspect the weapon for serviceability.

    a. Upper receiver group.

(1) Check handguards or rails for cracks, broken tabs, proper installation, and loose heat shields.

(2) Check front sight post for straightness.

(3) Check depression of the front detent.

(4) Check compensator for looseness.

(5) Check barrel for straightness, cracks, burrs or looseness.

(6) Check charging handle for cracks, bends, or breaks.

(7) Check rear sight assembly for properly working windage and elevation adjustments.

(8) Ensure the short and long range sight spring holds the selected sight in place.

(9) Check gas tube for bends or retention to barrel.

    b. Bolt carrier group.

(1) Inspect bolt cam pin for cracking or chipping.

(2) Inspect firing pin for bends, cracks, and sharp or blunted tip.

Note: Bolts that contain pits in the firing pin hole need replacing.

(3) Inspect for missing or broken gas rings.

(4) Inspect bolt cam pin area for cracking or chipping.

(5) Inspect locking lugs for cracking or chipping.

(6) Inspect extractor assembly for missing extractor spring assembly with insert and for chipped or broken edges on the lip which engages the cartridge rim.

(7) Inspect firing pin retaining pin to determine if bent or badly worn.

(8) Inspect bolt carrier for loose bolt carrier key.

(9) Inspect for cracking or chipping in cam pin hole area.

    c. Lower receiver.

(1) Inspect buffer for cracks or damage.

(2) Inspect buffer spring for kinks.

(3) Inspect buttstock for broken buttplate or cracks.

(4) Inspect for bent or broken selector lever.

(5) Inspect rifle grips for cracks or damage.

(6) Inspect for broken or bent trigger.

(7) Visually inspect the inside parts of the lower receiver for broken or missing parts.

    d. Turn in weapons with unserviceable parts for maintenance.

5. Lubricate the weapon.

Note: Under all but the coldest arctic conditions, CLP is the lubricant to use on the weapon. Temperature between +10 degrees fahrenheit and -10 degrees fahrenheit, use either CLP or LAW. For -35 degrees fahrenheit or lower, use LAW only. Lightly lube means apply a film of lubricant barely visible to the eye. Generously lube means apply the lubricant heavily enough so that it can be spread with the finger.

    a. Upper receiver and carrying handle.

        (1) Lightly lubricate inside of upper receiver, bore, chamber, front sight, outer surfaces of barrel, and under the handguards.

        (2) Apply a drop or two of lubricant to the front sight detent.

            (a) Depress and apply two or three drops of CLP to the front sight detent.

            (b) Depress several times to work the lube into the spring.

        (3) Apply a drop or two of lubricant to both threaded studs.

            (a) Lightly lube the clamping bar and both round nuts.

            (b) Lightly lube the mating surface.

        (4) Apply one or two drops of lubricant to the adjustable rear sight.

        (5) Ensure that the lubricant is spread evenly in the rear sight by rotating the following parts.

            (a) Elevation screw shaft.

            (b) Elevation knob.

            (c) Windage knob.

            (d) Windage screw.

    b. Lower receiver group.

        (1) Lightly lube the inside and outside lower receiver extension, buffer, and action spring.

            (a) Lightly lube the inside buttstock assembly.

            (b) Generously lube the buttstock lock-release lever and retaining pin.

        (2) Generously lube the take down pin, pivot pin, detents, and all other moving parts and their pins.

    c. Bolt carrier group.

        (1) Lightly lube the charging handle and the inner and outer surfaces of the bolt carrier.

        (2) Place one drop of CLP in the carrier key.

        (3) Apply a light coat of CLP on the firing pin and firing pin recess in the bolt.

        (4) Generously lube the outside of the bolt body, bolt rings, and cam pin area.

        (5) Apply a light coat of CLP on the extractor and pin.

6. Assemble the weapon.

    a. Install the buttstock assembly. (M4 series only)

        (1) Align the buttstock assembly with the lower receiver extension.

        (2) Pull downward on the lock release lever near the retaining pin.

        (3) Slide the buttstock assembly onto the lower receiver extension.

    b. Insert the action spring and buffer.

    c. Assemble the bolt carrier.

        (1) Insert the extractor and spring.

        (2) Push in the extractor pin.

(3) Slide the bolt into the carrier.

> **DANGER**
> The cam pin must be installed in the bolt group. Failure to do so will cause weapon to explode when fired next. Injury or death may occur.

    (4) Replace the bolt cam pin.

    (5) Drop in and seat the firing pin.

    (6) Pull the bolt back.

    (7) Replace the retaining pin.

  d. Engage and then push the charging handle in part of the way.

  e. Slide in the bolt carrier assembly.

  f. Push in the charging handle and the bolt carrier group together.

  g. Join the upper and lower receivers.

  h. Engage the receiver pivot pin.

  i. Close the upper and lower receiver groups.

  j. Push in the take down pin.

  k. Replace the handguards.

  l. Replace the carrying handle, if applicable.

  m. Replace the sling.

7. Perform a function check on the weapon.

8. Maintain the magazine.

  a. Disassemble magazine.

    (1) Insert the nose of a cartridge into the hole in the base of the magazine.

    (2) Raise the rear of the magazine until the indentation on the base is clear of the magazine.

    (3) Slide the base forward until it is free of the tabs.

    (4) Remove the magazine spring and follower (do not separate).

  b. Clean all parts using a rag soaked with CLP.

  c. Dry all parts.

  d. Inspect parts for damage such as dents and corrosion.

Note: If any damage is found, turn in to maintenance.

  e. Lightly lube the spring only.

    (1) Insert the follower and spring into the magazine tube.

    (2) Jiggle the spring to seat them in the magazine.

    (3) Slide the base under all four tabs until it is fully seated.

    (4) Make sure the printing is on the outside.

  f. Assemble the magazine.

9. Maintain the ammunition.

  a. Clean the ammunition with a clean dry rag.

  b. Inspect for and turn in any ammunition with the following defects:

    (1) Corrosion.

    (2) Dented cartridges.

    (3) Cartridges with loose bullets.

    (4) Cartridges with the bullet pushed in.

Evaluation Preparation:

Setup: Provide the Soldier with the equipment and or materials described in the conditions statement.

Brief Soldier: Tell the Soldier what is expected of him by reviewing the task standards. Stress to the Soldier the importance of observing all cautions, warnings, and dangers to avoid injury to personnel and, if applicable, damage to equipment.

Performance Measures	GO	NO GO
1 Cleared the weapon.	_____	_____
2 Disassembled the weapon.	_____	_____
3 Cleaned the weapon.	_____	_____
4 Inspected the weapon for serviceability.	_____	_____
5 Lubricated the weapon.	_____	_____
6 Assembled the weapon.	_____	_____
7 Performed a function check on the weapon.	_____	_____
8 Maintained the magazine.	_____	_____
9 Maintained the ammunition.	_____	_____

Evaluation Guidance: Score the Soldier GO if all performance measures are passed. Score the Soldier NO-GO if any performance measure is failed. If the

Soldier scores a NO-GO, show the Soldier what was done wrong and how to do it correctly.

References:
Required: FM 3-22.9, TM 9-1005-319-10
Related:

Environment: Environmental protection is not just the law but the right thing to do. It is a continual process and starts with deliberate planning. Always be alert to ways to protect our environment during training and missions. In doing so, you will contribute to the sustainment of our training resources while protecting people and the environment from harmful effects. Refer to FM 3-34.5 Environmental Considerations and GTA 05-08-002 ENVIRONMENTAL-RELATED RISK ASSESSMENT.

Safety: In a training environment, leaders must perform a risk assessment in accordance with ATP 5-19, Risk Management. Leaders will complete the current Deliberate Risk Assessment Worksheet in accordance with the TRADOC Safety Officer during the planning and completion of each task and sub-task by assessing mission, enemy, terrain and weather, troops and support available-time available and civil considerations, (METT-TC). Note: During MOPP training, leaders must ensure personnel are monitored for potential heat injury. Local policies and procedures must be followed during times of increased heat category in order to avoid heat related injury. Consider the MOPP work/rest cycles and water replacement guidelines IAW FM 3-11.4, Multiservice Tactics, Techniques, and Procedures for Nuclear, Biological, and Chemical (NBC) Protection, FM 3-11.5, Multiservice Tactics, Techniques, and Procedures for Chemical, Biological, Radiological, and Nuclear Decontamination.

071-COM-0029

Perform a Function Check on an M16-Series Rifle/M4-Series Carbine.

WARNING

Before starting functional check, be sure to clear the weapon. DO NOT squeeze the trigger until the weapon has been cleared. Inspect the chamber to ensure that it is empty and no ammunition is in position to be chambered.

Conditions: You are a member of a squad or team preparing for an tactical operation and must ensure the operability of your assigned M16-series rifle or M4-series carbine.

Standards: Conduct a function check and ensure that the weapon operates properly with the selector switch in each position.

Condition: None

Special Standards: None

Safety Risk: Low

Cue: None

*Note:*A function check is the final step of maintaining your weapon. It is also performed anytime the proper operation of a weapon is in question. Stop a function check at anytime the weapon does not function properly and turn in the malfunctioning weapon as per unit Standing Operating Procedures.

Performance Steps
1. Confirm the M16/M4 is clear.
2. Conduct a function check on the M16/M4.
   a. Place selector lever on SAFE.
   b. Pull charging handle to rear and release.
   c. Pull trigger.
Note: Hammer should not fall.
   d. Place selector lever on SEMI.
   e. Pull trigger.
Note: Hammer should fall.
   f. Hold trigger to the rear and charge the weapon.
   g. Release the trigger with a slow, smooth motion, until the trigger is fully forward.
Note: An audible click should be heard.
   h. Pull trigger.
Note: Hammer should fall.
   i. Place selector lever on BURST (M16A2, M16A4, and M4 only).
   j. Charge weapon one time.
   k. Squeeze trigger.
Note: Hammer should fall.
   l. Hold trigger to the rear.
   m. Charge weapon three times.
   n. Release trigger.
   o. Squeeze trigger.
Note: Hammer should fall.
   p. Place the selector switch on AUTO (M16A3 and M4A1 only).
   q. Pull the charging handle to the rear, charging the weapon.
   r. Squeeze the trigger.
Note: Hammer should fall.

s. Hold the trigger to the rear.
t. Cock the weapon again.
u. Fully release the trigger then squeeze it again.

Note: The hammer should not fall because it should have fallen when the bolt was allowed to move forward during the chambering and locking sequences.

Evaluation Preparation:

Setup: Provide the Soldier with the equipment and/or materials described in the conditions statement.

Brief Soldier: Tell the Soldier what is expected by reviewing the task standards. Stress to the Soldier the importance of observing all cautions, warnings, and dangers to avoid injury to personnel and, if applicable, damage to equipment.

Performance Measures	GO	NO GO
1 Confirmed the M16/M4 was clear.	_____	_____
2 Conducted a function check on the M16/M4.	_____	_____

Evaluation Guidance: Score the Soldier GO if all performance measures are passed. Score the Soldier NO-GO if any performance measure is failed. If the Soldier scores a NO-GO, show the Soldier what was done wrong and how to do it correctly.

References:
Required:
Related: FM 3-22.9, TM 9-1005-319-10

071-COM-0028

Load an M16-Series Rifle / M4-Series Carbine

Conditions: You are assigned a M16 series rifle or M4 carbine with magazines loaded with 5.56-mm ammunition. You must load it in preparation for operation. Some iterations of this task should be performed in MOPP 4.

Standards: Keep the weapon pointed in a safe direction, ensure chanmber is empty, place weapon on safe, insert a magazine, and chamber a round.

Special Condition: This task is being superseded by 071-COM-0028, REFER TO DTMS OR CAR (https://atiam.train.army.mil/catalog/catalog/search.html)for the new task

Special Standards: None

Special Equipment:

Cue: None

Note: None

Performance Steps

1. Determine the mode in which the weapon will be operated.
   a. Semiautomatic mode.
   b. Automatic mode for M16A1, M16A3, and M4A1.
   c. Burst mode for M16A2, M16A4, M4, and M4 - modular weapon system (MWS).
2. Point the weapon in a safe direction.
3. Cock the weapon.
   a. Pull the charging handle to the rear.
   b. Check the chamber to ensure it is clear.
Note: The chamber can be checked either by locking the bolt to the rear or by holding the bolt to the rear and then observing the chamber area.
   c. Return the charging handle to the forward position.
4. Place the selector lever on :
   a. SAFE for semiautomatic or automatic fire modes.
   b. BURST for burst fire mode.
5. Select BURST fire mode.
Note: Step 5 should only be performed if you have selected to fire the M16A2, M16A4, M4, or the M4 - MWS in the burst mode. For all other modes and weapons proceed to step 6.
   a. Rotate the BURST cam to the BURST position.
   b. Ensure the bolt is forward and the selector level is on BURST.
   c. Squeeze the trigger and hold it in the rear position.
   d. Pull the charging handle to the rear and release three times.
   e. Pull the charging handle to the rear one more time and hold it to the rear.
   f. Release the trigger.
   g. Lock the bolt open by pressing the bottom portion of the bolt catch.
   h. Return the charging handle to the forward position.
   i. Place the selector lever on SAFE.

6. Insert the magazine.

a. Push the magazine upwards until the magazine catch engages.

b. Tap upward on the bottom of the magazine to ensure the magazine is seated.

7. Chamber a round.

Note: A round may be chambered with the bolt assembly open or closed.

a. With the bolt open:

(1) Press the upper portion of the bolt catch allowing the bolt to go forward.

(2) Tap the forward assist to ensure that the bolt is fully forward and locked.

b. With the bolt closed :

(1) Pull the charging handle to the rear as far as it will go.

(2) Release the charging handle.

Note: The charging handle should not be rode forward.

WARNING
The weapon is now loaded and should be pointed in a safe direction.

(3) Tap the forward assist to ensure that the bolt is fully forward and locked.

8. Place selector lever on SAFE and close the ejection port cover if the weapon is not to be fired immediately.

Evaluation Preparation:

Setup: Have Soldiers use their assigned weapons and magazines. Provide blank or dummy ammunition.

Brief Soldier: Tell the Soldier to load the carbine.

Performance Measures	GO	NO GO
1 Keep the weapon pointed in a safe direction.	_____	_____
2 Ensured the chamber was clear.	_____	_____
3 Place the weapon on safe.	_____	_____
4 Locked the bolt to the rear.	_____	_____

Performance Measures	GO	NO GO
5 Inserted a magazine.	_____	_____
6 Chambered a round.	_____	_____

Evaluation Guidance: Score the Soldier GO if all performance measures are passed. Score the Soldier NO-GO if any performance measure is failed. If the Soldier scores NO-GO, show the Soldier what was done wrong and how to do it correctly.

References:
Required: FM 3-22.9, TM 9-1005-319-10
Related:

071-COM-0027

Unload an M16-Series Rifle / M4-Series Carbine

Conditions: You have just returned from a mission and have been directed to unload your M16-series rifle or M4-series carbine.Some iterations of this task should be performed in MOPP 4.

Standards: Unload the M16-series rifle or M4-series carbine so that the magazine and all ammunition are removed from the weapon.

Special Condition: None

Special Standards: None

Safety Risk: Medium

MOPP 4: Sometimes

Cue: None

Note: None

1. Point the weapon muzzle in a safe direction.
2. Place the selector lever on SAFE.

Note: If the weapon is not cocked, you cannot place the selector lever on SAFE.

3. Remove the magazine.
4. Lock the bolt open.
   a. Pull the charging handle to the rear.
   b. Press the bottom portion of the bolt catch, locking the bolt open.
   c. Return the charging handle to the forward position.
   d. Place the selector lever on SAFE.

Note: If the weapon was cocked before locking the bolt open then the selector lever should already be on SAFE.

5. Ensure that no ammunition is in the receiver and chamber.
6. Return the bolt to the closed position.
   a. Press the upper portion of the bolt catch allowing the bolt to go forward.
   b. Place selector lever on SEMI.
   c. Pull the trigger to release the pressure on the firing pin spring.
   d. Close the ejection port cover.

Evaluation Preparation:

Setup: At a test site, provide an M4 or M4A1 carbine loaded with dummy ammunition.

Brief Soldier: Tell the Soldier to unload the carbine.

Performance Measures	GO	NO GO
1 Pointed the weapon muzzle in a safe direction.	_____	_____
2 Placed the selector lever on SAFE.	_____	_____
3 Removed the magazine.	_____	_____
4 Locked the bolt open.	_____	_____

Performance Measures	GO	NO GO
5 Ensured no ammunition was in the receiver and chamber.	_____	_____
6 Returned the bolt to the closed position.	_____	_____

Evaluation Guidance: Score the Soldier GO if all performance measures are passed. Score the Soldier NO-GO if any performance measure is failed. If the Soldier scores NO-GO, show the Soldier what was done wrong and how to do it correctly.

References
Required: FM 3-22.9, TM 9-1005-319-10
Related:

071-COM-0030

Engage Targets with an M16-Series Rifle/M4 Series Carbine

Conditions: You are a member of a squad conducting dismounted operations and have been assigned a sector of fire by your leader. You have your M16 series rifle or M4 series carbine, magazines, ammunition, and individual combat/personal protective equipment. Some iterations of this task should be performed in MOPP 4.
Standards: Select a firing position and engage targets in your assigned sector until they no longer present a threat or you are directed to cease fire.

Performance Steps

1. Select a position that allows for adequate observation of assigned sector of fi
Note: Your situation should affect your physical positioning and firing stance. Your position should protect you from enemy fire and observation, yet allow you to place effective fire on targets in your sector of fire. Your position may varyfrom a fixed location to a temporary location during movement.
Note: Detection of targets depends on your position, your skill in scanning, and your ability to observe the are and recognize target indicators.
2. Scan sector of fire using one of the following methods.
   a. Self preservation method

   b.50-meter overlapping strip method

   c.Maintaining observation on the area
3. Identify targets in designated sector of fire.

4. Determine range to targets

  a. 100-meter unit of measure method

  b. Appearance of objects method.

  c. Front sight post method

  d. Appearance of objects method.
  e. Combination method.

5. Fire on targets using correct fundamentals of marksmanship and appropriate aiming and engagement techniques
  a. Apply the fundamentals of marksmanship.
  (1) Steady position

  (2) Aiming.
  (3) Breath control.
  (4) Trigger squeeze.
  b. Use appropriate aiming and engagement techniques as needed.
  (1) Combat fire techniques.
  (2) Chemical, biological, radiological and nuclear (CBRN) firing.
  (3) Night firing.
  (4) Moving targets.
  (5) Short-range marksmanship techniques.
  (6) Cease fire on targets once they are destroyed, suppressed, or you receive an order to cease fire.

Evaluation Preparation: *Setup:* On a live-fire range, provide sufficient quantities of equipment and ammunition to support the number of Soldiers tested. Have each Soldier use his own rifle and magazine.

Brief Soldier: Tell Soldier that he/she is to detect and engage targets in his/her sector and, when asked, state the range to the target.

At a test site, provide an M4 or M4A1 carbine loaded with dummy ammunition.

	Performance Measures	**GO**	**NO GO**
1	Selected a position that allowed for adequate observation of assigned sector of fire.	____	____
2	Scanned sector of fire.	____	____

Performance Measures	GO	NO GO
3 Identified target in designated sector of fire.	_____	_____
4 Determined range to target.	_____	_____
5 Fired on targets.	_____	_____
6. Ceased fire once targets were destroyed, suppressed, or you were directed to cease fire .	_____	_____

Evaluation Guidance: Score the Soldier GO if all performance measures are passed. Score the Soldier NO-GO if any performance measure is failed. If the Soldier scores a NO-GO, show the Soldier what was done wrong and how to do it correctly.

References:
Required: FM 3-22.9, TC 3-21.75, TM 9-1005-319-10
Related:

071-COM-0033

Correct Malfunctions of an M16-Series Rifle / M4-Series Carbine

Conditions: You have a stoppage while engaging targets with your M16-series rifle or M4-series carbine. Some iterations of this task should be performed in MOPP 4.

Standards: Perform immediate and/or remedial action so you can continue to engage targets.

Special Condition: None

Ssfety Risk: Medium

Special Equipment:

Cue: None

Note: None

1. Perform immediate action.
Note: The key word "SPORTS" will help you remember the steps for immediate action in sequence; slap, pull, observe, release, tap, shoot.

WARNING
The weapon is now loaded and should be pointed in a safe direction.

    a. Slap upward on the magazine to ensure it is fully seated and that the magazine follower is not jammed..
Note: When slapping up on the magazine, be careful not to knock a round out of the magazine into the line of the bolt carrier.
    b. Pull the charging handle fully to the rear.
    c. Observe the ejection of a live round or expended cartridge.
Note: If a weapon fails to eject a cartridge, perform remedial action.
    d. Release the charging handle; do not ride the charging handle.
    e. Tap the forward assist to ensure that the bolt is closed.
    f. Squeeze the trigger and try to fire the rifle.

DANGER

If weapon stops firing with a live round in the chamber of a hot barrel, remove the round quickly. However, if you cannot remove it within 10 seconds, remove magazine and wait 15 minutes with the weapon pointed in a safe direction. This will avoid injury during possible cook-off. Always keep face away from the ejection port when clearing a hot chamber.

Note: Apply immediate action only once for a stoppage. If the rifle fails to fire a second time for the same malfunction remedial action should be performed.
2. Perform remedial action.
    a. Correct an obstructed chamber.
        (1) Lock the charging handle to the rear.
        (2) Place the weapon on SAFE.
        (3) Remove the magazine.

(4) Visually inspect the chamber.**DANGER**

DO NOT attempt to remove a round stuck in the barrel of a weapon; turn the weapon in to field maintenance.

        (5) Remove obstructions from the chamber by:
            (a) Angling the ejection port downward and shaking the rifle to remove single rounds.
            (b) Using a pointed object to lessen jammed rounds then shake out when loose.
            (c) Using a cleaning rod to push out a round or cartridge case stuck in the chamber.

b. Correct a mechanical malfunction.
(1) Clear the weapon.
(2) Disassemble the weapon.
(3) Inspect for dirty, corroded, missing, or broken parts.
(4) Clean dirty or corroded parts.
(5) Replace missing or broken parts.
(6) Assemble the weapon.
(7) Perform a function check.

Evaluation Preparation:

Setup: Provide an M4 or M4A1 carbine loaded with dummy ammunition.

Brief Soldier: Tell the Soldier that the rifle has stopped firing. Tell the Soldier that the weapon is cool and that he/she is to perform the immediate or remedial actions on the rifle. All steps must be performed in the proper sequence.

Performance Measures	GO	NO GO
1 Performed immediate action.	_____	_____
2 Performed remedial action.	_____	_____

Evaluation Guidance: Score the Soldier GO if all performance measures are passed. Score the Soldier NO-GO if any performance measure is failed. If the Soldier scores NO-GO, show the Soldier what was done wrong and how to do it correctly.

References:
Required: FM 3-22.9, TM 9-1005-319-10
Related:

071-COM-0031

Zero an M16 Series Rifle / M4-Series Carbine

Conditions: You are assigned an M16-series rifle or M4-series carbine and have been directed to zero the weapon. You have 18 rounds of 5.56-mm ammunition, the appropriate 25-meter zero target, and sandbags for support. Some iterations of this task should be performed in MOPP 4.

Standards: Fire the weapon and adjust the sights so that five out of six rounds in two consecutive shot groups strike within the 4-centimeter circle on the target using 18 rounds or less. Record your zero.

Special Condition: None

Special Standards: None

Special Equipment:

Cue: None

Note: None

Performance Steps

1. Set either the battlesight zero or mechanical zero on your weapon.
   a. Determine whether to set a mechanical zero or the battlesight zero.
      (1) Set a mechanical zero if-
         (a) The weapon sights have been serviced.
         (b) The weapon is newly assigned to the unit.
         (c) The current zero on the weapon is questionable.
      (2) Set a battlesight zero if a mechanical zero is not required.
   b. Set a mechanical zero on your weapon.
      (1) Adjust the Front Sight.
         (a) Move the front sightpost until the base of the front sightpost is flush with the front sightpost housing.
         (b) (M16A1 only) Move the front sightpost, from the flush position, 11 clicks in the direction of UP.
      (2) Adjust the Rear Sight (by weapon type).
         (a) (M16A1 only) Turn the rear sight windage drum left until it stops.
         (b) (M16A1 only) Turn the windage drum right 17 clicks to center it.
         (c) (M16A2 / M16A3 / M16A4 / M4-series) Set rear apertures by positioning the apertures so the unmarked aperture is up and the 0-200 meter aperture is down.
         (d) (M16A2 / M16A3 / M16A4 / M4-series) Set windage by turning the windage knob to align the index mark on the 0-200 meter aperture with the long center index line on the rear sight assembly.
         (e) (M16A2 / M16A3) Set the elevation of the M16A2/A3 by turning the elevation knob counterclockwise until the rear sight assembly rests flush with the carrying handle and the 8 / 3 marking is aligned with the index line on the left side of the carrying handle.
         (f) (M16A4 only) Turn the elevation knob counterclockwise until the rear sight assembly rests flush with the carrying handle and the 6 / 3 marking is aligned with the index line on the left side of the carrying handle.

(g) (M4-series only) Turn the elevation knob counterclockwise until the rear sight assembly rests flush with the detachable carrying handle and the 6 / 3 marking is aligned with the index line on the left side of the carrying handle.

c. Set a battlesight zero on your weapon.

Note: No changes are made to the front sight when setting a battlesight zero.

(1) (M16A1 only) Adjust Rear Sight by flipping the aperture to ensure the aperture marked "L" is visible.

(2) (M16A2 / M16A3 / M16A4 / M4-Series only) Adjust rear aperture by positioning the apertures so the unmarked aperture is up and the 0-200 meter aperture is down.

(3) (M16A2 / M16A3 / M16A4 / M4-Series only) Adjust windage by turning the windage knob to align the index mark on the 0-200 meter aperture with the long center index line on the rear sight assembly.

(4) (M16A2 / M16A3 only) Adjust elevation by-

(a) Turning the elevation knob counterclockwise until the rear sight assembly rests flush with the carrying handle and the 8 / 3 marking is aligned with the index line on the left side of the carrying handle.

(b) Turning the elevation knob one more click clockwise.

(5) (M16A4 only) Adjust elevation by-

(a) Turning the elevation knob counterclockwise until the rear sight assembly rests flush with the carrying handle and the 6 / 3 marking is aligned with the index line on the left side of the carrying handle.

(b) Turning the elevation knob two more clicks clockwise so the index line on the left side of the detachable carrying handle is aligned with the "Z" on the elevation knob.

(6) (M4-series only) Adjust elevation by turning the elevation knob counterclockwise until the rear sight assembly rests flush with the detachable carrying handle and the 6 / 3 marking is aligned with the index line on the left side of the carrying handle.

2. Establish a correct sight picture.

a. Confirm the correct 25-meter zero target is facing you.

b. Assume a prone supported firing position.

c. Align the sights.

(1) Center the top of the front sight post in the center of the rear sight.

(2) Visualize imaginary cross hairs in the center of the rear aperture so that the top of the front sight post touches the imaginary horizontal line and the front sight post bisects imaginary vertical line.

(3) Verify the sight picture.

d. Align the aiming point.

(1) Aim at target center.

(2) Position the top of the front sight post center mass of the scaled silhouette target.

(3) Confirm that an imaginary vertical line drawn through the center of the front sight post splits the target.

(4) Confirm that an imaginary horizontal line drawn through the top of the front sight post splits the target.

3. Establish a tight shot group.

Note: A tight shot group is 3 consecutive rounds within a 4 centimeter or less circle.

 a. Fire a three round shot group at the 25-meter zeroing target.

 b. Identify the shot group on the target.

 c. Repeat step 3a and step 3b until 2 consecutive 3 round shot groups fall within a 4 centimeter or less circle.

Note: If a tight shot group is not obtained after 18 rounds then remedial training must be done.

 4. Adjust sights (if required) to obtain a zero.

Note: Do not adjust the sights your just fired shot groups meet the standard.

 a. Determine the necessary sight adjustments by identifying the center of the last fired shot group and identifying the adjustment to move this point to the center of the strike zone (zero offset).

Note: The numbered squares around the edges of the target each represent a click on the sight.

 b. Adjust Elevation.

Note: One click clockwise moves the strike of the bullet down one square, while one click counterclockwise moves the strike of the bullet up one square.

  (1) Find the horizontal line nearest the center of the shot group.

  (2) Follow the line either left or right to the nearest edge of the target.

  (3) Identify the number of clicks and the direction of adjustment shown at the edge of the target.

  (4) Adjust the front sight in the indicated direction by the appropriate number of clicks.

  (5) Record the adjustment made on the target.

 c. Adjust Windage.

Note: Three clicks counterclockwise moves the strike of the bullet left one square, while three clicks clockwise moves the strike of the bullet right one square.

  (1) Find the vertical line (up and down) nearest the center of the shot group.

  (2) Follow the line either up or down to the nearest edge of the target.

  (3) Identify the number of clicks and the direction of adjustment shown at the edge of the target.

  (4) Adjust the rear sight in the indicated direction by the appropriate number of clicks.

  (5) Record the adjustment made on the target.

 5. Establish a zero.

 a. Fire a three round shot group at the 25-meter zeroing target.

 b. Identify the location of the shot group on the target.

  (1) Return to step 4, if 2 of 3 rounds do not strike within the strike zone / zero offset.

  (2) Proceed to step 6 if 2 of 3 rounds strike within the strike zone / zero offset.

 6. Confirm the zero.

Note: A zero is confirmed when 5 of 6 rounds land within the center 4 centimeter center circle or the zero offset circle.

    a. Fire a three round shot group at the 25-meter zeroing target.

    b. Identify the location of the shot group on the target.

        (1) Return to step 4, if 2 of 3 rounds do not strike within the strike zone / zero offset.

        (2) Cease fire if 2 of 3 rounds strike within the strike zone / zero offset (your zero is confirmed).

    7. (M4-series only) Rotate the rear sight elevation knob counterclockwise (down) two clicks to the 300-meter setting.

    8. Record your zero.

    a. Compute your zero.

    b. Write your zero on a piece of tape.

    c. Attach the tape to your weapon.

Evaluation Preparation:

Setup: Provide the Soldier with the equipment and or materials described in the conditions statement.

Brief Soldier: Tell the Soldier what is expected of him by reviewing the task standards. Stress to the Soldier the importance of observing all cautions, warnings, and dangers to avoid injury to personnel and, if applicable, damage to equipment.

Performance Measures	GO	NO GO
1 Set either the mechanical zero or the battlesight zero on your weapon.	_____	_____
2 Established a correct sight picture.	_____	_____
3 Established a tight shot group.	_____	_____
4 Adjusted sights (if required) to obtain a zero.	_____	_____
5 Established a zero.	_____	_____
6 Confirmed the zero.	_____	_____

Performance Measures	GO	NO GO
7 (M4-series only) Rotated the rear sight elevation knob counterclockwise (down) two clicks to the 300-meter setting.	_____	_____
8 Recorded your zero.	_____	_____

Evaluation Guidance: Score the Soldier GO if all performance measures are passed. Score the Soldier NO-GO if any performance measure is failed. If the Soldier scores a NO-GO, show the Soldier what was done wrong and how to do it correctly.

References
Required: FM 3-22.9, TM 9-1005-319-10
Related:

071-COM-4401

Perform Safety Checks on Hand Grenades

Conditions: You are a member of a squad or team preparing for a mission and have been directed to perform safety checks on the hand grenades issued to your squad/team. The hand grenades are in a shipping container. You are wearing your individual combat/personal protective equipment. Some iterations of this task should be performed in MOPP 4.

Standards: Inspect the shipping container, canister, and hand grenade for defects; report and turn in hand grenade that has defect(s) that cannot be corrected; secure hand grenade(s) properly in carrying pouch(s).

Special Condition: None

Safety Risk: Medium

MOPP 4: Somrtimes

Cue: None

*Note:*If any discrepancies are found upon receipt of an issued shipping container, canister or hand grenade, personnel should return the shipping container, canister or hand grenade to the issuing person or dispose of it in accordance with the unit tactical standing operating procedures(TACSOP).

Performance Steps

1. Inspect hand grenade shipping container (Figure 071-COM-4401-1), if applicable.

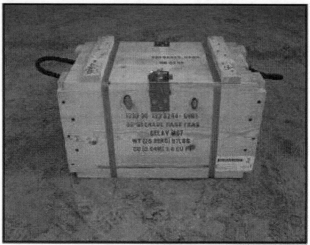

Figure 071-COM-4401-1 Shipping container.

    a. Shipping container is not damaged.
    b. Inform supervisor if shipping container is damaged.
2. Inspect the grenade canister (Figure 071-COM-4401-2), if applicable.

Figure 071-COM-4401-2. Grenade canister.

a. Inspect the canister for damage.

(1) Check to see if seal on the canister has been tampered with or is missing.

(2) Ensure canister is not dented or punctured.

(3) Inform supervisor of any deficiencies found.

b. Open the canister.

WARNING

Do not attempt to remove the grenade found upside down in its packing container.

---

(1) Check to see if the grenade is upside down inside of the shipping canister.

(a) Replace canister top and tape in place if grenade found upside down.

(b) Report deficiencies to supervisor.

(c) Return canister to ammunition disposal personnel.

(2) Check to see if the safety pin is in proper position.

(a) Ensure that safety pin is in place and undamaged.

(b) Check that the legs of the safety pin have either angular spread or diamond crimp.

(3) Ensure safety clip (when installed) is in place and undamaged.

3. Inspect the hand grenade.

a. Remove the packing material and the hand grenade from the canister.

(1) Check for rust on the body or the fuze.

(2) Ensure holes are not visible in the body or the fuze.

(3) Check hand grenade for cracked body.

(4) Place back in canister if any defect(s) are found, if applicable.

b. Ensure the safety pin (1) is secured properly (Figure 071-325-4401-3).

Note: If not properly secured, carefully push it into place while holding the safety lever down.

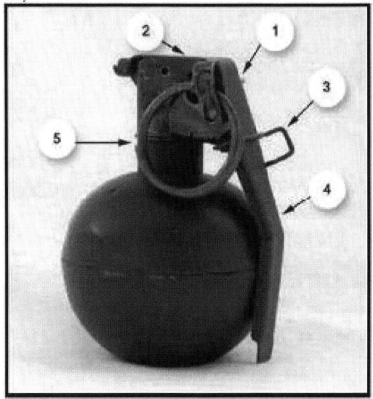

Figure 071-COM-4401-3. Grenade components.

    c. Ensure the confidence clip (2) is present and properly secured to the pull ring.

    d. Ensure the safety clip (3) is present and properly secured to the safety lever (4).

Note: If not properly secured, carefully push it into place while holding the safety lever down.

WARNING

Never remove the fuze from a live grenade.

---

    e. Check the hand grenade fuze assembly (5) for tightness.
    f. Ensure the safety lever (4) is not bent or broken.
    g. Turn in defective hand grenade, if applicable.
  4. Secure the grenade.

WARNING

Never carry the grenades suspended by the safety pull ring or safety lever. Do not attach grenades to clothing or equipment by the pull ring. Do not tape hand grenades to Soldier's gear. Do not attempt to modify a grenade.

---

    a. Carry hand grenades using the proper procedures.
    b. Ensure that the grenade is fully inside the carrying pouch.
    c. Secure pouch flap.

Evaluation Preparation:

Setup: Provide the Soldier with the equipment and or materials described in the conditions statement.

Brief Soldier: Tell the Soldier what is expected of him by reviewing the task standards. Stress to the Soldier the importance of observing all cautions, warnings, and dangers to avoid injury to personnel and, if applicable, damage to equipment.

Performance Measures	GO	NO GO
1 Inspected hand grenade shipping container, if applicable.	____	____
2 Inspected the hand grenade canister, if applicable.	____	____
3 Inspected the hand grenade.	____	____
4 Secured the hand grenade.	____	____
5 Report any deficiencies to Supervisor.	____	____

Evaluation Guidance: Score the Soldier GO if all performance measures are passed. Score the Soldier NO-GO if any performance measure is failed. If the Soldier scores a NO-GO, show the Soldier what was done wrong and how to do it correctly.

References:
Required: TC 3-23-30; TM 9-1330-200-12
Related:

071-COM-4407

Employ Hand Grenades

Conditions: Given a fragmentation, concussion, riot control, smoke, or incendiary grenade with a time-delay fuse, a point or area target to engage, and load bearing vest (LBV), load bearing equipment (LBE), Modular, Lightweight, Load-bearing, Equipment (MOLLE) or Improved Outer Tactical Vest (IOTV). Some iterations of this task should be performed in MOPP 4.

Standards: Engage target with a hand grenade by: selecting appropriate hand grenade based on type target, determining throwing position, correctly gripping, preparing, and throwing the hand grenade so it is within the effective range of the target.

Special Condition: None

Special Standards: None

Safety Risk: High

Cue: None

MOPP 4: Sometimes

Note: None

Performance Steps

   1. Select appropriate hand grenade based on type of target.
   2. Select proper throwing position.
Note: You can use five positions to throw grenades - standing, prone-to-standing, kneeling, prone-to-kneeling, and alternate prone. However, If you can

achieve more distance and accuracy using your own personal style, do so as long as your body is facing sideways and toward the enemy's position, and you throws the grenade overhand.

    a. Ensure you have a proper covered position.

    b. Determine the distance to the target.

    c. Align your body with the target.

  3. Grip the hand grenade.

Note: Do not remove the safety clip or the safety pin until the grenade is about to be thrown.

    a. Place the hand grenade in the palm of the throwing hand with the safety lever placed between the first and second joints of the thumb.

Note: For left handed throwers the grenade is inverted with the top of the fuze facing downwards in the throwing hand.

    b. Keep the pull ring away from the palm of the throwing hand so that it can be easily removed by the index or middle finger of the free hand.

  4. Prepare the hand grenade.

    a. Tilt the grenade forward to observe the safety clip.

    b. Remove the safety clip by sweeping it away from the grenade with the thumb of the opposite hand.

    c. Insert the index or middle finger of the nonthrowing hand in the pull ring until it reaches the knuckle of the finger (Figure 071-COM-4407-1).

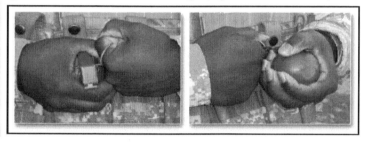

Figure 071-COM-4407-1. Pull ring grip, right/left hand.

DANGER

If pressure on the safety lever is relaxed after the safety clip pin are removed, the striker can rotate and strike the primer while the thrower is still holding the grenade. Continuing to hold the grenade beyond this point can result in injury or death.

---

    d. Ensure that you are holding the safety lever down firmly.

    e. Twist the pull ring toward the body (away from the body for left handed throwers) to release the pull ring from the confidence clip.

CAUTION

Never attempt to reinsert a safety pin into a hand grenade during training. In combat, however, it may be necessary to reinsert a safety pin into a grenade. Take special care to replace the pin properly. If the tactical situation allows, it is safer to throw the grenade rather than to trust the reinserted pin.

     f. Remove the safety pin by pulling the pull ring from the grenade (Figure 071-COM-4407-2 and Figure 071-COM-4407-3).

Figure 071-COM-4407-2. Right hand grip, pulling safety pin.

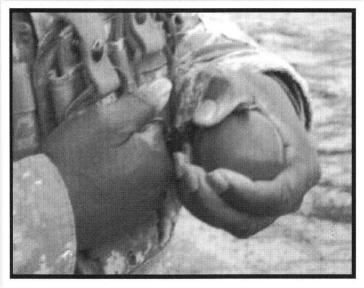

Figure 071-COM-4407-3. Left hand grip, pulling the safety pin.

5. Throw the hand grenade so it is within the effective range of the target.

a. Observe the target to estimate the distance between the throwing position and the target area.

Note: In observing the target, minimize exposure time to the enemy (no more than 3 seconds).

WARNING
The flight path of the grenade must be checked to make sure no obstacles alter the flight of the grenade or cause it to bounce back toward you.

b. Ensure there are no obstacles that can alter or block the flight of the grenade when it is thrown.

c. Confirm body target alignment.

DANGER
Use cook-off procedures only in a combat environment. In training, never cook off live fragmentation hand grenades or offensive concussion grenades. Never cook off the M84, stun grenade, or smoke grenades. These grenades have short fuze delays (1 to 2.3 seconds) and will cause serious personal injury if cook-off procedures are performed. The grenade must be thrown immediately after count off.

    d. Cook off the hand grenade. (Optional)

Note: Cooking off uses enough of the grenade's 4- to 5-second delay (about 2 seconds) to cause the grenade to detonate above ground or shortly after impact with the target.

    (1) Release the safety lever.

    (2) Count "One thousand one, one thousand two".

    e. Throw the grenade overhand so that the grenade arcs, landing on or near the target.

Note: To be effective the target must be within the bursting radius of the grenade.

    f. Allow the motion of the throwing arm to continue naturally once the grenade is released.

    g. Seek cover to avoid being hit by fragments or direct enemy fire.

Note: If no cover is available, drop to the prone position with your protective head gear facing the direction of the grenade's detonation.

Evaluation Preparation:

Setup: Provide the Soldier with the equipment and or materials described in the conditions statement.

Brief Soldier: Tell the Soldier what is expected of him by reviewing the task standards. Stress to the Soldier the importance of observing all cautions, warnings, and dangers to avoid injury to personnel and, if applicable, damage to equipment.

	Performance Measures	GO	NO GO
1	Selected the appropriate hand grenade based on type of target.	_____	_____
2	Selected appropriate throwing position.	_____	_____
3	Gripped the hand grenade.	_____	_____
4	Prepared the grenade.	_____	_____
5	Threw the hand grenade so it was within the effective range of the target.	_____	_____

Performance Measures	GO	NO GO

Evaluation Guidance: Score the Soldier GO if all performance measures are passed. Score the Soldier NO-GO if any performance measure is failed. If the Soldier scores a NO-GO, show the Soldier what was done wrong and how to do it correctly.

References:
Required:
Related: TC 3-23.30, TM 9-1330-200-12

Subject Area 2: Move

071-COM-0541

Perform Exterior Movement Techniques during an Urban Operation

Conditions: You are a member of a dismounted squad or team conducting movement within an urban area. You have your assigned weapon and individual/protective equipment. The enemy's location and strength in the area are unknown.

Standards: Move within an urban area using proper urban movement techniques while minimizing exposure to enemy fire.

Special Condition: None

Special Standards: None

Special Equipment:

Cue: None

Note: Outdoor movement in urban terrain is best conducted as part of a buddy team, a fire team, or a squad. This ensures at least one Soldier is providing overwatch of another Soldier's movement, either from a stationary position or as both are moving, and prevents individual Soldiers from being isolated. This allows for a rapid engagement of any enemy that either exposes themselves (such as by leaning out of or by silhouetting themselves in a window) or by firing.

1. Move across streets or open areas.

Note: Open areas include parks, plazas and large intersections as well as streets, open air buildings, and large rooms that are significantly exposed to exterior view. Ideally, avoid these open areas as they are potential killing zones for the enemy, especially crew-served weapons and snipers; however, operations often require movement across these areas. Cross these areas using the same basic techniques used to cross any danger area.

    a. Identify the far side position before moving with a clear understanding of how it will be occupied or cleared.

    b. Conduct a visual reconnaissance of all the dimensions of urban terrain to identify likely threat positions.

    c. Select a position on the far side that provides the best available cover.

    d. Select the best route to the far side position that minimizes the time exposed.

Note: Obscurants, such as smoke, are an option to conceal movement. However, thermal sighting systems can see through smoke and when smoke is thrown in an open area, an enemy may fire into the smoke cloud in anticipation of movement through or behind the smoke.

    e. Cross rapidly along the selected route to the selected position.

2. Move parallel to buildings.

Note: Moving parallel to buildings is the movement normally associated with moving down a roadway but also includes movement in plazas or other open areas that are between buildings. During contact, utilize smoke, suppressive fires, and individual movement techniques. In moving to adjacent buildings, team members should keep a distance of 3 to 5 meters between themselves, leapfrogging along each side of the street and from cover to cover.

    a. Soldier moves parallel to the side of a building.

    b. Use existing cover and concealment.

    c. Stay in the shadows.

    d. Present a low silhouette.

    e. Use proper techniques to cross door and window openings

    f. Move rapidly to the next position.

3. Move past building opening (windows and doors).

Note: The most common mistakes at windows are exposing the head in a first-floor window and not being aware of basement windows.

    a. Move past an above-knee window.

      (1) Stay near the side of the building.

      (2) Stay below the level of the window.

      (3) Avoid silhouetting self in window (Figure 071-COM-0541-1).

Figure 071-COM-0541-1. Soldier moving past window.

    b. Move past a below-knee window (basement).
      (1) Stay near the side of the building.
      (2) Step or jump past the window without exposing legs (Figure 071-COM-0541-2).

Figure 071-COM-0541-2. Soldier moving past basement window.

c. Move past a full-height window (store type) or open door.

Note: A Soldier should not just walk past an adjacent full height window, as he presents a good target to an enemy inside the building.

(1) Identify a position on the far side of the window.

(2) Determine which technique to use to cross the opening.

(a) Run across the opening to the far side.

(b) Arc around the opening while covering the opening with your weapon while moving.

(3) Move rapidly to the far side position.

4. Move around corners.

Note: Before moving around a corner, the Solider must first observe around the corner. The most common two mistakes Soldier make at corners are exposing their head and upper body where it is expected and flagging their weapon.

a. Move around a corner by first observing around the corner.

(1) Lie flat on the ground, weapon at your side, ensuring that your weapon is not forward of the corner.

Note: DO not show your head below at the height an enemy would expect to see it.

(2) Expose your head (with Helmet) only enough to observe around the corner (Figure 071-COM-0541-3).

Note: When speed is required the Pie-ing method is applied.

Figure 071-COM-0541-3. Soldier looking around a corner.

(3) Continue movement around the corner, if clear.

b. Move around the corner by using the pie-ing method.

(1) Aim the weapon beyond the corner (without flagging) into the direction of travel.

(2) Side-step around the corner in a semi-circular fashion with the muzzle as the pivot point (Figure 071-COM-0541-4).

Figure 071-COM-0541-4. Soldier Pie-ing around a corner.

(3) Continue movement around the corner, if clear.

5. Cross a wall.

a. Reconnoiter the other side.

Note: The far side must be relatively safe from enemy fire, as once across the wall, the Soldier is fully exposed. Additionally, the immediate opposite side of the wall must be safe for landing; long drops and debris can cause injury.

b. Identify a far side position.

Note: Once across the wall, you will then move to this far side position. This position may be at the wall, near the wall, or away from the wall.

c. Crouch near the wall.

d. Hold your weapon with one hand while grabbing the top of the wall with the other hand

e. Pull with the hand on the wall while simultaneously swinging both legs over the wall, one right after the other.

f. Roll your whole body quickly over the wall while keeping a low silhouette (Figure 071-COM-0541-5).

Note: Speed of movement and a low silhouette deny the enemy a good target.

Figure 071-COM-0541-5. Soldier crossing a wall.

g. Move to you next position once on the far side.

Evaluation Preparation:

Setup: At the test site, provide all materials and equipment given in the task condition statement.

Brief Soldier: Tell the Soldier to move as a designated member of an assult element in urban terrain. The enemy strength and location are unknown.

Performance Measures	GO	NO GO
1 Moved across a street or open area.	_____	_____
2 Moved parallel to a building.	_____	_____
3 Moved passed a building opening (window or open door).	_____	_____

Performance Measures	GO	NO GO
4 Moved around a corner.	_____	_____
5 Crossed a wall.	_____	_____

Evaluation Guidance: Score the Soldier GO if all performance measures are passed. Score the Soldier NO-GO if any performance measure is failed. If the Soldier scores NO-GO, show the Soldier what was done wrong and how to do it correctly.

References:
Required: ATTP 3-06.11; TC 3-21.75
Related:

071-COM-0503

Move Over, Through, or Around Obstacles (Except Minefields)

Conditions: As a member of a dismounted team conducting movement to contact, you encounter a natural or manmade obstacle. You have your assigned weapon and individual/protective equipment. The enemy's location and strength in the area are unknown. Some iterations of this task should be performed in MOPP 4.

Standards: Notify chain of command of obstacle encountered, evaluate obstacle, identify nearest covered position on far side of obstacle, negotiate a wall obstacle, and provide local security for follow on forces during engotiation or reduction of obstacle.

Special Condition: None

Special Standards: None

Safety Risk: Medium

MOPP 4: Sometimes

Cue: None

*Note:*An obstacle is any obstruction designed or employed to disrupt, fix, turn, or block the movement of an opposing force, and to impose additional losses in personnel, time, and equipment on the opposing force. Obstacles can be natural, manmade, or a combination of both.

Performance Steps

1. Notify your chain of command of the presence and type of obstacle encountered.

Note: Most obstacles, for maximum effectiveness, are covered by either fire or observation. Many obstacles, due to enemy fire or complexity of the obstacle, require a unit breaching operation and the appropriate collective task should also be followed.

2. Evaluate the obstacle, from a covered position, to determine whether to move around, through or over the obstacle

Note: Typically it is best to move around (or bypass) an obstacle, however this is not always possible.

3. Identify the nearest covered position on the far side of the obstacle.

4. Ensure a buddy, if present, covers your movement as you negotiate the obstacle.

5. Negotiate a wall obstacle.

    a. Identify your immediate landing position on the far side of the wall.

Note: The far side must be relatively safe from enemy fire, as once across the wall, you are fully exposed. Additionally, the immediate opposite side of the wall must be safe for landing as long drops and debris can cause injury.

    b. Assume a crouching position near the wall, while holding your weapon with one hand and grabbing the top of the wall with the other hand.

    c. Pull with the hand on the wall while simultaneously swinging both legs over the wall, one right after the other.

    d. Roll quickly over the top to other side, keeping a low silhouette.

    e. Move to the identified covered position on the far side.

WARNING

An enemy may attach booby traps or tripwire-activated mines to wire obstacles.

---

6. Negotiate a wire obstacle.

    a. Move to your designated crossing position.

    b. Check for booby traps or early warning devices.

    c. Cross over a wire obstacle.

      (1) Place an object such as a piece of wood, metal, or mats, over the wire.

      (2) Move over the wire by stepping on this object to avoid the wire entanglements.

    d. Cross under a wire obstacle.

      (1) Slide head first on your back under the bottom strands.

      (2) Push yourself forward with your shoulders and heels, carrying your weapon lengthwise on your body and holding the barbed wire with one hand while moving.

      (3) Let the barbed wire slide on the weapon to keep wire from catching on clothing and equipment.

e. Cut through a wire obstacle.

Note: If stealth is not needed then quickly cut all wires and proceed through the gap.

    (1) Wrap cloth around the barbed wire between your hands.

    (2) Cut partly through the barbed wire.

Note: Cutting the wire near a picket reduces the noise of a cut.

    (3) Bend the barbed wire back and forth quietly until it separates.

    (4) Cut only the lower strands.

    (5) Cross under the remaining top wires.

7. Cross a ditch type obstacle.

    a. Select a point that has cover and concealment on both sides, such as a bend in the ditch.

    b. Move to your designated crossing site.

    c. Crawl up to the edge of the open area.

    d. Observe both the floor of the ditch and the far side for dangers.

    e. Move rapidly but quietly across the exposed area.

    f. Assume a covered position on the far side.

8. Cover your buddy, if present, as he or she crosses the obstacle.

Evaluation Preparation:

Setup: Provide the Soldier with the equipment and or materials described in the conditions statement.

Brief Soldier: Tell the Soldier what is expected of him by reviewing the task standards. Stress to the Soldier the importance of observing all cautions, warnings, and dangers to avoid injury to personnel and, if applicable, damage to equipment.

Performance Measures	GO	NO GO
1 Notified the chain of command of the presence and type of obstacle encountered.	_____	_____
2 Evaluated the obstacle, from a covered position, to determine whether to move around, through or over the obstacle.	_____	_____
3 Identified the nearest covered position on the far side of the obstacle.	_____	_____

Performance Measures	GO	NO GO
4 Ensured team members, if present, provide local security for your movements as you negotiated the obstacle.	_____	_____
5 Negotiated a wall obstacle.	_____	_____
6 Provide local security on far side of obstacles for follow on forces, if present.	_____	_____

Evaluation Guidance: Score the Soldier GO if all performance measures are passed. Score the Soldier NO-GO if any performance measure is failed. If the Soldier scores a NO-GO, show him what was done wrong and how to do it correctly.

References
Required: TC 3-21.75
Related:

071-COM-1000

Identify Topographic Symbols on a Military Map

Conditions: You are a member of a squad or team in a field environment and have been given; a 1:50,000 scale military map and a requirement to identify topographic symbols on the map.

Standards: Identify topographic symbols, colors, and marginal information on a military map.

Special Condition: None

Special Standards: None

Safety Risk: None

MOPP 4:

Cue: None

Note: None

1. Identify the six basic colors on a military map (Figure 071-COM-1000-1).

COLORS	SYMBOLS
Black	Cultural (man-made) features other than roads
Blue	Water
Brown	All relief features - contour lines on old maps - cultivated land on red-light readable maps
Green	Vegetation
Red	Major roads, built-up areas, special features on old maps
Red-brown	All relief features and main roads on red-light readable maps

Figure 071-COM-1000-1. Colors

    a. Identify the features that the color black represents.
Note: Indicates cultural (manmade) features such as buildings and roads, surveyed spot elevations, and all labels.
    b. Identify the features that the color blue represents.
Note: Indicates hydrography or water features such as lakes, swamps, rivers, and drainage.
    c. Identify the features that the color green represents.
Note: Indicates vegetation with military significance such as woods, orchards, and vineyards.
    d. Identify the features that the color brown represents.
Note: Brown identifies all relief features and elevation such as contours on older edition maps and cultivated land on red light readable maps.
    e. Identify the features that the color red represents.

Note: Classifies cultural features, such as populated areas, main roads, and boundaries, on older maps.

    f. Identify the features that the color red-brown represents.

Note: These colors are combined to identify cultural features, all relief features, non surveyed spot elevations, and elevation such as contour lines on red light readable maps.

    g. Identify all other features and the colors they represent, if applicable.

Note: Other colors may be used to show special information. These are indicated in the marginal information as a rule.

    2. Identify the symbols on a military map.

    a. Use the legend, which should identify most of the symbols used on the map.

    b. Identify each object by its shape on the map.

Note: For example, a black, solid square represents a building or a house; a round or irregular blue item is a lake or pond.

    c. Use logic and color to identify each map feature.

Note: For example, blue represents water. If you see a symbol that is blue and has clumps of grass, this would be a swamp.

    3. Identify the marginal information on a military map (Figure 071-COM-1000-2).

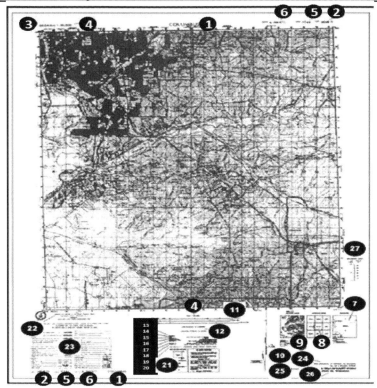

Figure 071-COM-1000-2. Topographical map.

a. Identify the sheet name (1).
b. Identify the sheet number (2).
c. Identify the series name (3).
d. Identify the scale (4).
e. Identify the series number (5).
f. Identify the edition number (6).
g. Identify the index to boundaries (7).
h. Identify the adjoining sheets diagram (8).
i. Identify the elevation guide (9).
j. Identify the declination diagram (10).
k. Identify the bar scales (11).
l. Identify the contour interval note (12).
m. Identify the spheroid note (13).
n. Identify the grid note (14).
o. Identify the projection note (15).
p. Identify the vertical datum note (16).
q. Identify the horizontal datum note (17).

r. Identify the control note (18).
s. Identify the preparation note (19).
t. Identify the printing note (20).
u. Identify the grid reference box (21).
v. Identify the unit imprint and symbol (22).
w. Identify the legend (23).

Evaluation Preparation:

Setup: Provide the Soldier with the equipment and or materials described in the conditions statement.

Brief Soldier: Tell the Soldier what is expected of him by reviewing the task standards. Stress to the Soldier the importance of observing all cautions, warnings, and dangers to avoid injury to personnel and, if applicable, damage to equipment.

Performance Measures	GO	NO GO
1 Identified the six basic colors on a military map.	____	____
2 Identified the symbols on a military map.	____	____
3 Identified the marginal information on a military map.	____	____

Evaluation Guidance: Score the Soldier GO if all performance measures are passed. Score the Soldier NO-GO if any performance measure is failed. If the Soldier scores a NO-GO, show the Soldier what was done wrong and how to do it correctly.

References:
Required:
Related: TC 3-25.26

071-COM-1001

Identify Terrain Features on a Map

Conditions: You are a member of a squad or team in a field environment and have been directed to identify the terrain features on a map. You have been given a 1:50,000 scale military map.

Standards: Identify the five major, three minor, and two supplementary terrain features on a military map.

Special Condition: None

Safety Risk: Low

MOPP 4:

Cue: None

Note: All terrain features are derived from a complex landmass known as a mountain or ridgeline (Figure 071-COM-1001-1). The term ridgeline is not interchangeable with the term ridge. A ridgeline is a line of high ground, usually with changes in elevation along its top and low ground on all sides from which a total of 10 natural or man-made terrain features are classified.

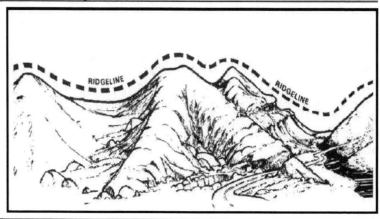

Figure 071-COM-1001-1. Ridgeline.

1. Identify five major terrain features.
    a. Identify a hill (Figure 071-COM-1001-2).
Note: A hill is an area of high ground. From a hilltop, the ground slopes down in all directions. A hill is shown on a map by contour lines forming concentric circles. The inside of the smallest closed circle is the hilltop.

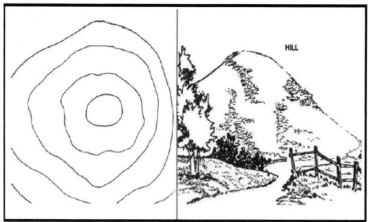

Figure 071-COM-1001-2. Hill.

    b. Identify a saddle (Figure 071-COM-1001-3).
Note: A saddle is a dip or low point between two areas of higher ground. A saddle is not necessarily the lower ground between two hilltops; it may be simply a dip or break along a level ridge crest. If you are in a saddle, there is high ground in two opposite directions and lower ground in the other two

directions. A saddle is normally represented as an hourglass.

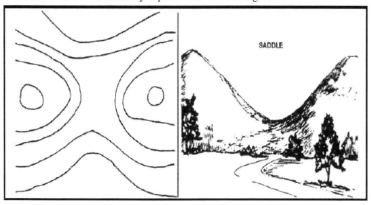

Figure 071-COM-1001-3. Saddle.

    c. Identify a valley (Figure 071-COM-1001-4).

Note: A valley is a stretched-out groove in the land, usually formed by streams or rivers. A valley begins with high ground on three sides and usually has a course of running water through it. If standing in a valley, three directions offer high ground, while the fourth direction offers low ground. Depending on its size and where a person is standing, it may not be obvious that there is high ground in the third direction, but water flows from higher to lower ground. Contour lines forming a valley are either U-shaped or V-shaped. To determine the direction water is flowing, look at the contour lines. The closed end of the contour line (U or V) always points upstream or toward high ground.

Figure 071-COM-1001-4. Valley.

d. Identify a ridge (Figure 071-COM-1001-5).

Note: A ridge is a sloping line of high ground. If you are standing on the centerline of a ridge, you will normally have low ground in three directions and high ground in one direction with varying degrees of slope. If you cross a ridge at right angles, you will climb steeply to the crest and then descend steeply to the base. When you move along the path of the ridge, depending on the geographic location, there may be either an almost unnoticeable slope or a very obvious incline. Contour lines forming a ridge tend to be U-shaped or V-shaped. The closed end of the contour line points away from high ground.

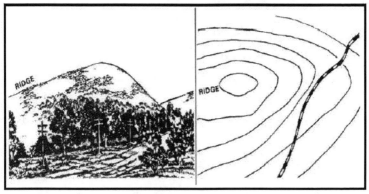

Figure 071-COM-1001-5. Ridge.

e. Identify a depression (Figure 071-COM-1001-6).

Note: A depression is a low point in the ground or a sinkhole. It could be described as an area of low ground surrounded by higher ground in all directions, or simply a hole in the ground. Usually only depressions that are equal to or greater than the contour interval will be shown. On maps, depressions are represented by closed contour lines that have tick marks pointing toward low ground.

Figure 071-COM-1001-6. Depression.

2. Identify three minor terrain features.
   a. Identify a draw (Figure 071-COM-1001-7).
Note: A draw is a stream course that is less developed than a valley. In a draw, there is essentially no level ground and, therefore, little or no maneuver room within its confines. If you are standing in a draw, the ground slopes upward in three directions and downward in the other direction. A draw could be considered as the initial formation of a valley. The contour lines depicting a draw are U-shaped or V-shaped, pointing toward high ground.

Figure 071-COM-1001-7. Draw.

   b. Identify a spur (Figure 071-COM-1001-8).
Note: A spur is a short, continuous sloping line of higher ground, normally jutting out from the side of a ridge. A spur is often formed by two roughly parallel streams cutting draws down the side of a ridge. The ground will slope

down in three directions and up in one. Contour lines on a map depict a spur with the U or V pointing away from high ground.

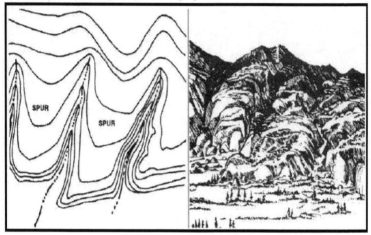

Figure 071-COM-1001-8. Spur.

   c. Identify a cliff (Figure 071-COM-1001-9).

Note: A cliff is a vertical or near vertical feature; it is an abrupt change of the land. When a slope is so steep that the contour lines converge into one "carrying" contour of contours, this last contour line has tick marks pointing toward low ground. Cliffs re also shown by contour lines very close together and, in some instances, touching each other.

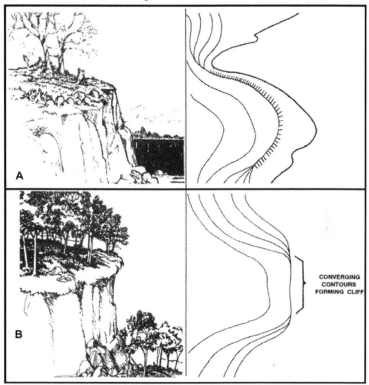

Figure 071-COM-1001-9. Cliff.

3. Identify two supplementary terrain features.

a. Identify a cut (Figure 071-COM-1001-10).

Note: A cut is a man-made feature resulting from cutting through raised ground, usually to form a level bed for a road or railroad track. Cuts are shown on a map when they are at least 10 feet high, and they are drawn with a contour line along the cut line. This contour line extends the length of the cut and has tick marks that extend from the cut line to the roadbed, if the map scale permits this level of detail.

b. Identify a fill (Figure 071-COM-1001-10).

Note: A fill is a man-made feature resulting from filling a low area, usually to form a level bed for a road or railroad track. Fills are shown on a map when they are at least 10 feet high, and they are drawn with a contour line along the fill line. This contour line extends the length of the filled area and has tick marks that point toward lower ground. If the map scale permits , the length of the fill tick marks are drawn to scale and extend from the base line of the fill symbol.

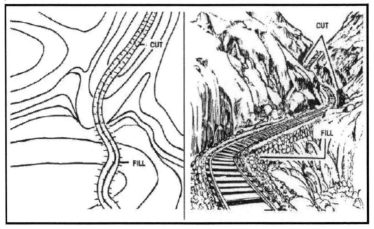

Figure 071-COM-1001-10. Cut and Fill.

Evaluation Preparation:

Setup: Provide the Soldier with the equipment and or materials described in the conditions statement.

Brief Soldier: Tell the Soldier what is expected of him by reviewing the task standards. Stress to the Soldier the importance of observing all cautions, warnings, and dangers to avoid injury to personnel and, if applicable, damage to equipment.

Performance Measures	GO	NO GO
1 Identified the five major terrain features.		
2 Identified the three minor terrain features.		
3 Identified the two supplementary terrain features.		

Evaluation Guidance: Score the Soldier GO if all performance measures are passed. Score the Soldier NO-GO if any performance measure is failed. If the Soldier scores a NO-GO, show the Soldier what was done wrong and how to do it correctly. n

References:
Required:
Related: TC 3-25.26

071-COM-1008

Measure Distance on a Map

Conditions: You are a member of a squad or team in a field environment and have been directed to determine the distance between two known points. You have a 1:50,000 scale map, a strip of paper with a straight edge, and a pencil. You have been shown the beginning and ending points on the map. Some iterations of this task should be performed in MOPP 4.

Standards: Determine the straight-line distance between two points with no more than a 5 percent error and the road (curved line) distance between two points with no more than a 10 percent error.

Special Condition: None

Special Standards: None

Safety Risk: Low

MOPP 4: Sometimes

Cue: None

Note: None

Performance Steps

1. Identify the graphic (bar) scale of the map.
2. Determine straight-line distance between two points on a map.
   a. Line up the straight edge of a strip of paper with the beginning and ending points on the map.
   b. Mark the beginning and ending points on the straight edge of the paper (Figure 071-COM-1008-1).

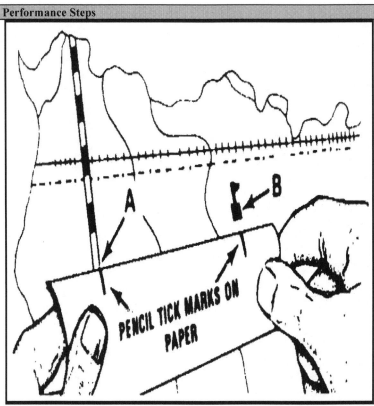

Figure 071-COM-1008-1. Beginning and Ending Points.

    c. Place the starting point on the paper under the zero on the bar scale.

    d. Measure off 4,000 meters and place a new tick mark on the paper.

    e. Place the new tick mark under the zero on the bar scale.

    f. Determine if the end point falls within the bar scale.

       (1) Record the value on the scale of the end point, if the end point fits on the scale.

       (2) Add 4,000 meters to this value (a) to get the total difference.

    g. Determine if the end point falls outside the bar.

       (1) Repeat steps 3d and 3e until the end point falls within the bar.

       (2) Add 4,000 meters to the value you derived in step 3f(1) for each time you performed step 3d to achieve the total distance.

    3. Convert map distance to ground distance.

    a. Align the edge of a strip of paper with the beginning point and the point where the road makes the first curve on the map.

    b. Mark on the straight edge of the paper the beginning and curve points.

c. Repeat steps 4a and b, each time using the point of the curve as the next beginning point, until you reach the end point.

d. Align the marks on the paper with the appropriate bar scale (Figure 071-COM-1008-2).

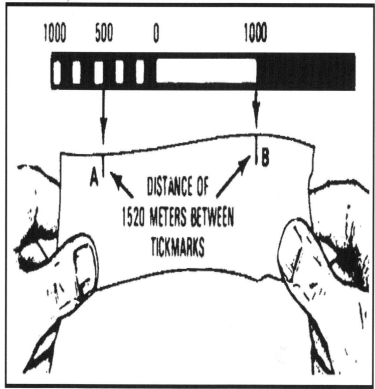

Figure 071-COM-1008-2. Distance between Beginning and Ending Points.

e. Determine the distance on the scale that compares to the distance on the paper.

4. Convert a road map distance to miles, meters or yards.

a. Align the edge of a strip of paper with the beginning point and the point where the road makes the first curve on the map.

b. Mark on the straight edge of the paper the beginning and curve points.

c. Repeat steps 5a and b, each time using the point of the curve as the next beginning point, until you reach the end point.

d. Place the starting point on the paper under the zero on the bar scale.

e. Measure off 4,000 meters and place a new tick mark on the paper.

f. Place the new tick mark under the zero on the bar scale.

g. Determine if the end point falls within the bar scale.

(1) Record the value on the scale of the end point, if the end point fits on the scale.

(2) Add 4,000 meters to this value (a) to get the total difference.

h. Determine if the end point falls outside the bar.

(1) Repeat steps 5d and 5e until the end point falls within the bar.

(2) Add 4,000 meters to the value you derived in step 5g(1) for each time you performed step 5d to achieve the total distance.

Evaluation Preparation:

Setup: Provide the Soldier with the equipment and or materials described in the conditions statement.

Brief Soldier: Tell the Soldier what is expected of him by reviewing the task standards. Stress to the Soldier the importance of observing all cautions, warnings, and dangers to avoid injury to personnel and, if applicable, damage to equipment.

Performance Measures	GO	NO GO
1 Identified the scale of the map.	_____	_____
2 Determine the straight-line distance between two points on a map.	_____	_____
3 Measure the distance along a road, stream, or other curved line.	_____	_____

Evaluation Guidance: Score the Soldier GO if all performance measures are passed. Score the Soldier NO-GO if any performance measure is failed. If the Soldier scores a NO-GO, show the Soldier what was done wrong and how to do it correctly.

References:
Required:
Related: TC 3-25.26

071-COM-1002

Determine the Grid Coordinates of a Point on a Military Map

Conditions: You are a member of a squad or team in a field environment and have been directed to identify the grid coordinates of a point on a map. You have a 1:50,000 scale military map, a coordinate scale and protractor or plotting scale, a pencil, and paper. You have been shown the point on the map. Some iterations of this task should be performed in MOPP 4.

Standards: Determine the coordinates of the grid square, determine grid coordinates of a point with and without a coordinate scale and protractor or plotting scale. Identify the 100,000 meter square identifier to determine grid coordinate.

Special Condition: None

Special Standards: None

Safety Risk: Low

MOPP 4: Sometimes

Cue: None

Note: None

Performance Steps

1. Determine the coordinates of the grid square (Figure 071-COM-1002-1).

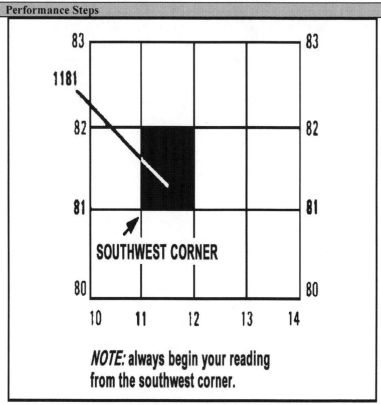

Figure 071-COM-1002-1. Identifying the Grid Square

a. Select the grid square that contains the identified point on the map (see Figure 071-COM-1002-1).

b. Read the north-south grid line that precedes the desired point (see Figure 071-COM-1002-1).

c. Record the number associated with that line.

d. Read the east-west grid line that precedes the desired point (see Figure 071-COM-1002- 1).

e. Record the number associated with that line.

Note: The number of digits represents the degree of precision to which a point has been located and measured on a map the more digits the more precise the measurement. In the above example the four digits 1181 identify the 1,000 meter grid square to be used.

2. Determine point grid coordinates without a coordinate scale and protractor or plotting scale (Figure 071-COM-1002-2).

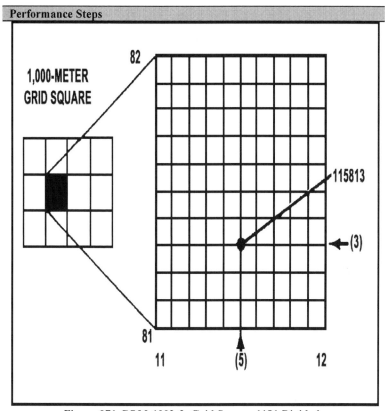

Figure 071-COM-1002-2. Grid Square 1181 Divided.

a. Allocate the grid square into a 10 by 10 grid.

b. Read right (from the lower left corner) to the imaginary gird line nearest the identified point.

Note: In the example the North-South imaginary line nearest the point is halfway or 5 lines out of a total of 10 lines. Therefore the first half of your grid coordinate is 115.

c. Read up (from the point reached in step 3b) to the imaginary gird line nearest the identified point.

Note: In the example the East-West imaginary line nearest the point is one third of the way up or 3 lines out of 10 lines. Therefore the second half of your grid coordinate is 813.

3. Determine point grid coordinates with coordinate scale and protractor or plotting scale (Figure 071-COM-1002-3).

Note: The most accurate way to determine the coordinates of a point on a map is with a coordinate scale. You need not imagine lines, because you can find the exact coordinates using the coordinate scale, protractor or the plotting scale.

Each device actually includes two coordinate scales, 1:25,000 and 1:50,000 meters. Make sure that, regardless which device you use, you choose the correct scale.

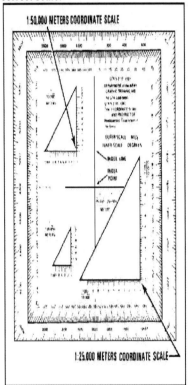 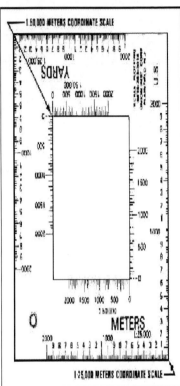

Figure 071-COM-1002-3. Coordinate Scale and Protractor (Left) and Plotting scale (Right).

a. Locate the grid square where the point is located (Example: Point A in Figure 071-COM-1002-4).

b. Determine the coordinates of the grid square.

NoteE: The number of the vertical grid line on the left (west) side of the grid square gives the first and second digits of the coordinate. The number of the horizontal grid line on the bottom (south) side of the grid square gives the fourth and fifth digits of the coordinate.

c. Determine the third and sixth digits of the coordinate.

(1) Place a coordinate scale and protractor or a plotting scale on the bottom horizontal grid line of the grid square containing Point A.

(2) Check to see that the zeros of the coordinate scale are in the lower left-hand (southwest) corner of the grid square where Point A is located (Figure 071-COM-1002-4).

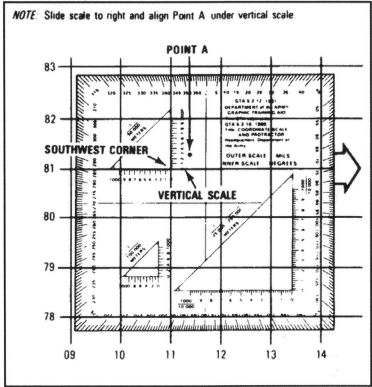

Figure 071-COM-1002-4. Placement of the Coordinate Scale.

(3) Slide the scale to the right, keeping the bottom of the scale on the bottom grid line until Point A is under the vertical (right-hand) scale (Figures 071-COM-1002-5 and 071-COM-1002-6).

Note: To determine the six-digit coordinate, look at the 100-meter mark on the bottom scale, which is nearest the vertical grid line. This mark is the third digit of the number 115. The 100-meter mark on the vertical scale nearest to Point A gives you the sixth digit of the number 813. The complete grid coordinate is 115813. Always read right, and then up.

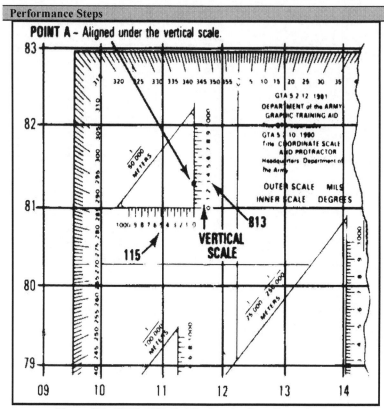

Figure 071-COM-1002-5. Aligning the Coordinate Scale.

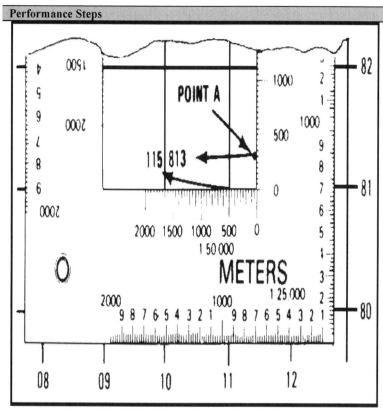

Figure 071-COM-1002-6. Aligning the Plotting Scale.

4. Add the two letter 100,000 meter square identifier to determined grid coordinate.

a. Identify the two letter 100,000 meter square identifier by looking at the grid reference box in the margin of the map (Figure 071-COM-1002-7).

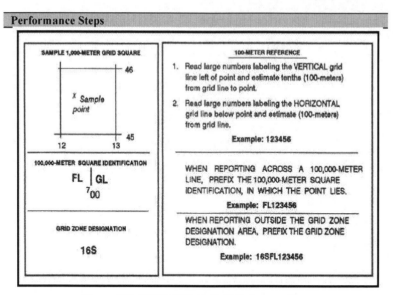

Figure 071-COM-1002-7. Grid Reference Box.

b. Place the 100,000 meter square identifier in front of the grid coordinate.
Note: In the example given the final grid coordinate becomes GL115813.
Evaluation Preparation:

Setup: Provide the Soldier with the equipment and or materials described in the conditions statement.

Brief Soldier: Tell the Soldier what is expected of him by reviewing the task standards. Stress to the Soldier the importance of observing all cautions, warnings, and dangers to avoid injury to personnel and, if applicable, damage to equipment.

Performance Measures	GO	NO GO
1 Determined the coordinates of the grid square.	_____	_____
2 Determined point grid coordinates without a coordinate scale and protractor or plotting scale.	_____	_____
3 Determined point grid coordinates with coordinate scale and protractor or plotting scale.	_____	_____

Performance Measures	GO	NO GO
4 Added the two letter 100,000 meter square identifier to the determined grid coordinate.	____	____

Evaluation Guidance: Score the Soldier GO if all performance measures are passed. Score the Soldier NO-GO if any performance measure is failed. If the Soldier scores a NO-GO, show the Soldier what was done wrong and how to do it correctly.

References:
Required:
Related: TC 3-25.26

071-COM-1005

Determine a Location on the Ground by Terrain Association

Conditions: You are a member of a squad or team in a field environment and have been directed to determine your squad's/team's current location. You have a 1:50,000 scale military map, a compass, a coordinate scale and protractor or plotting scale, a pencil, and paper. Some iterations of this task should be performed in MOPP 4.

Standards: Orient the map. Identify the type of terrain on which you are located as well as the type of terrain that surrounds your location. Correlate the terrain features on the ground to those shown on the map. Determine the six digit grid coordinates to your location..

Special Condition: None

Special Standards: None

Safety Risk: Low

MOPP 4: Sometimes

Cue: None

Note: None

Performance Steps
  1. Orient the map.

Note: There are three ways to orient a map:
- Using a compass. The magnetic arrow of the compass points to magnetic north. As such, pay special attention to the declination diagram.
- Using terrain association. This method is typically used when a compass is not available or when the user has to make many quick references as he moves across country.
- Using Field-Expedient Methods. These methods are used when a compass is available and there are no recognizable terrain features.
    2. Identify the type of terrain feature on which you are located.
    3. Identify the types of terrain features that surround your location.
    4. Correlate the terrain features on the ground to those shown on the map.
    5. Determine your location on the map.
    6. Determine the six digit grid coordinate of your location.
Note: Grid coordinates of your location can be determined by using a coordinate scale and protractor, a plotting scale, or by visualizing a 10 by 10 grid box inside the appropriate grid square.

Evaluation Preparation:

Setup: Provide the Soldier with the equipment and or materials described in the conditions statement.

Brief Soldier: Tell the Soldier what is expected of him by reviewing the task standards. Stress to the Soldier the importance of observing all cautions, warnings, and dangers to avoid injury to personnel and, if applicable, damage to equipment.

Performance Measures	GO	NO GO
1 Oriented the map.	_____	_____
2 Identified the type of terrain feature on which you were located.	_____	_____
3 Identified the types of terrain features that surround your location.	_____	_____
4 Correlated the terrain features on the ground to those shown on the map.	_____	_____

Performance Measures	GO	NO GO
5 Determined your location on the map.	_____	_____
6 Determined the six digit grid coordinate of your location.	_____	_____

Evaluation Guidance: Score the Soldier GO if all performance measures are passed. Score the Soldier NO-GO if any performance measure is failed. If the Soldier scores a NO-GO, show the Soldier what was done wrong and how to do it correctly.

References
Required:
Related: TC 3-25.26

071-COM-1012

Orient a Map to the Ground by Map-Terrain Association

Conditions: You are a member of a squad or team that is conducting movement in a field environment and you have been directed to orient a standard 1:50,000 scale military map to the ground. You do not have an operational compass. Some iterations of this task should be performed in MOPP 4.

Standards: Hold the map horizontally and match terrain features appearing on the map with physical features on the ground. Orient the map to within 30 degrees of magnetic north.

Special Condition: None

Special Standards: None

Safety Risk: Low

Cue: None

Note: A map can be oriented by terrain association when a compass is not available or when the user has to make many quick references as he moves across country. Using this method requires careful examination of the map and the ground, and the user must know his approximate location.

1. Hold the map in a horizontal position.

2. Match terrain features appearing on your map with terrain features physically observable on the ground (Figure 071-COM-1012-1).

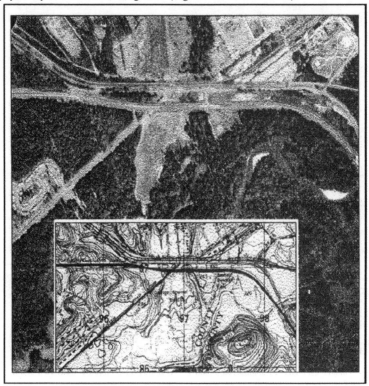

Figure 071-COM-1012-1. Terrain Association.

3. Align the map with the terrain features on the ground.

Evaluation Preparation:

Setup: Provide the Soldier with the equipment and or materials described in the conditions statement.

Brief Soldier: Tell the Soldier what is expected of him by reviewing the task standards. Stress to the Soldier the importance of observing all cautions, warnings, and dangers to avoid injury to personnel and, if applicable, damage to equipment.

Performance Measures	GO	NO GO
1 Held the map in a horizontal position.	____	____
2 Matched terrain features appearing on map with physical features on the ground.	____	____
3 Aligned the map with the terrain features on the ground to within 30 degrees of magnetic north.	____	____

Evaluation Guidance: Score the Soldier GO if all performance measures are passed. Score the Soldier NO-GO if any performance measure is failed. If the Soldier scores a NO-GO, show the Soldier what was done wrong and how to do it correctly.

References:
Required:
Related: TC 3-25.26

071-COM-1011

Orient a Map Using a Lensatic Compass

Conditions: You are a member of a squad or team in a field environment and have been directed to orient a map in preparation for movement. You have a 1:50,000-scale topographic map of the area and a compass. Some iterations of this task should be performed in MOPP 4.

Standards: Determine the direction and value of declination, lay the map in a horizontal position and orient the map to the ground using a compass.

Special Condition: None

Special Standards: None

Safety Risk: Low

Cue: None

Note: The first step for a navigator in the field is orienting the map. A map is oriented when it is in a horizontal position with its north and south corresponding to the north and south on the ground.

When orienting a map with a compass, remember that the compass measures magnetic azimuths. Since the magnetic arrow points to magnetic north, pay special attention to the declination diagram. Two techniques are used.

Special care should be taken when orienting your map with a compass. A small mistake can cause you to navigate in the wrong direction.

Once the map is oriented, magnetic azimuths are determined using the compass. Do not move the map from its oriented position since any change in its position moves it out of line with the magnetic north.

Performance Steps

1. Determine the direction of the declination and its value from the declination diagram on the map.
2. Lay the map in a horizontal position.
3. Use one of the two techniques to orient the map.
   a. Orient the map using the first technique.
   (1) Take the straightedge on the left side of the compass and place it alongside the north-south grid line with the cover of the compass pointing toward the top of the map.
Note: This procedure places the fixed black index line of the compass parallel to north-south grid lines of the map.
   (2) Keep the compass aligned as directed above while rotating the map and compass together until the magnetic arrow is below the fixed black index line on the compass.
Note: At this time, the map is close to being oriented.
   (3) Rotate the map and compass in the direction of the declination diagram.
   (4) Verify the G-M angle.
      (a) If the magnetic north arrow on the map is to the left of the grid north, check the compass reading to see if it equals the G-M angle given in the declination diagram (Figure 071-COM-1011-1).

Figure 071-COM-1011-1. Map oriented with 10 degrees west declination.

(b) If the magnetic north is to the right of grid north, check the compass reading to see if it equals 360 degrees minus the G-M angle (Figure 071-COM-1011-2).

Note: If the G-M angles are correct the map is oriented.

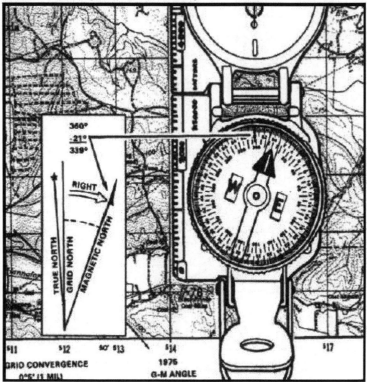

Figure 071-COM-1011-2. Map oriented with 21 degrees east declination.

    b. Orient the map using the second technique.

        (1) Draw a magnetic azimuth equal to the G-M angle given in the declination diagram with the protractor using any north-south grid line on the map as a base.

        (2) If the declination is easterly (right), the drawn line is equal to the value of the G-M angle:

            (a) Align the straightedge on the left side of the compass alongside the drawn line on the map.

            (b) Rotate the map and compass until the magnetic arrow of the compass is below the fixed black index line (Figure 071-COM-1011-3).

Note: The map is now oriented.

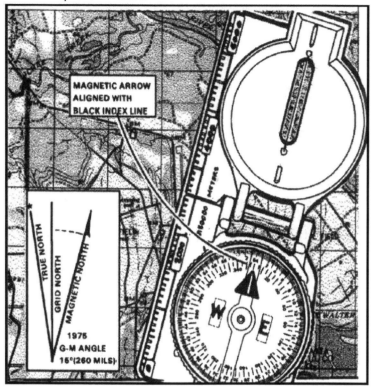

Figure 071-COM-1011-3. Map oriented with 15 degrees east declination.

(3) If the declination is westerly (left), the drawn line will equal 360 degrees minus the value of the G-M angle:

(a) Align the straightedge on the left side of the compass alongside the drawn line on the map.

(b) Rotate the map and compass until the magnetic arrow of the compass is below the fixed black index line.

Note: The map is now oriented.

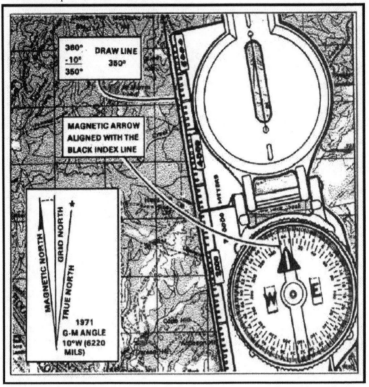

Figure 071-COM-1011-4. Map oriented with 10 degrees west declination.

Evaluation Preparation:

Setup: Provide the Soldier with the equipment and or materials described in the conditions statement.

Brief Soldier: Tell the Soldier what is expected of him by reviewing the task standards. Stress to the Soldier the importance of observing all cautions, warnings, and dangers to avoid injury to personnel and, if applicable, damage to equipment.

Performance Measures	GO	NO GO
1 Determined the direction of the declination and its value from the declination diagram.	_____	_____
2 Laid the map in a horizontal position.	_____	_____
3 Used one of the two techniques to orient the map.	_____	_____

Evaluation Guidance: Score the Soldier GO if all performance measures are passed. Score the Soldier NO-GO if any performance measure is failed. If the Soldier scores a NO-GO, show the Soldier what was done wrong and how to do it correctly.

References:
Required:
Related: TC 3-25.26

071-COM-1003

Determine a Magnetic Azimuth Using a Lensatic Compass

Conditions: You are a member of a squad or team in a field environment and have been directed to determine a magnetic azimuth. You have a compass and a designated point on the ground. Some iterations of this task should be performed in MOPP 4.

Standards: Inspect the compuss. Determine the correct magnetic azimuth to the designated point within 3 degrees using the compass-to-cheek method, and within 10 degrees using the center-hold method.

Special Condition: None

Special Standards: None

Safety Risk: Low

MOPP 4: Sometimes

Cue: None

1. Inspect the compass (Figure 071-COM-1003-1).

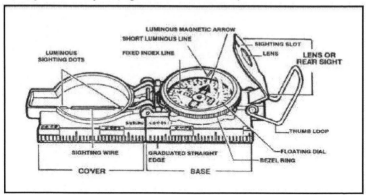

Figure 071-COM-1003-1. Lensatic compass.

    a. Ensure floating dial, which contains the magnetic needle moves freely and does not stick.

    b. Ensure the sighting wire is straight.

    c. Ensure glass and crystal parts are not broken.

    d. Ensure numbers on the dial are readable.

2. Determine direction (Figure 071-COM-1003-2).

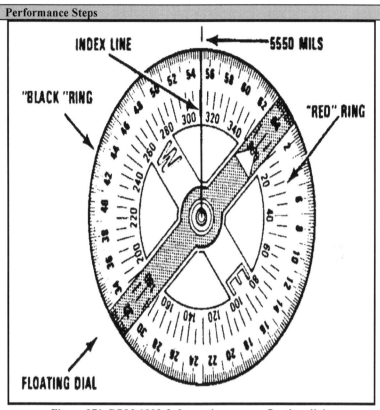

Figure 071-COM-1003-2. Lensatic compass floating dial.

    a. Align the compass to the direction you want to go or want to determine.

    b. Locate the scale beneath the index line on the outer glass cover.

    c. Determine to the nearest degree, or 10 mils, the position of the index line over the red or black scale.

Note: Effects of Metal and Electricity. Metal objects and electrical sources can affect the performance of a compass. However, nonmagnetic metals and alloys do not affect compass readings. The following separation distances are suggested to ensure proper functioning of a compass:

High-tension power lines .... 55 meters.
Field gun, truck, or tank... 18 meters.
Telegraph or telephone wires and barbed wire....... 10 meters.
Machine gun .... 2 meters.
Steel helmet or rifle... 1/2 meter.

   3. Determine an azimuth with the compass-to-cheek method (Figure 071-COM-1003-3).

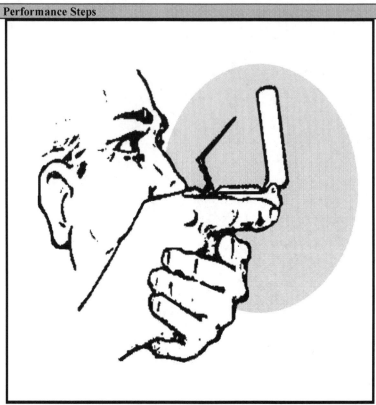

Figure 071-COM-1003-3. Compass-to-cheek method.

a. Open the cover to a 90-degree angle to the base.

b. Position the eyepiece at a 45-degree angle to the base.

c. Place your thumb through the thumb loop.

d. Establish a steady base with your third and fourth fingers.

e. Extend your index finger along the side of the compass base.

f. Place the hand holding the compass into the palm of the other hand.

g. Move both hands up to your face.

h. Position the thumb that is through the thumb loop against the cheekbone.

i. Move the eyepiece up or down until the dial is in focus.

j. Align the sighting slot of the eyepiece with the sighting wire in the cover on the desired point.

k. Read the azimuth under the index line.

4. Determine an azimuth with the center-hold method (Figure 071-COM-1003-4).

Note: This method offers the following advantages over the sighting technique:

-It is faster and easier to use.

-It can be used under all conditions of visibility.

-It can be used when navigating over any type of terrain.

-It can be used without putting down the rifle; however, the rifle must be slung well back over either shoulder.

-It can be used without removing eyeglasses

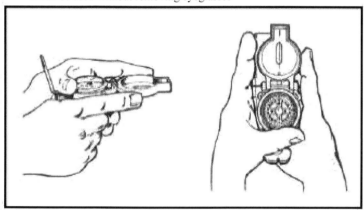

Figure 071-COM-1003-4. Centerhold technique.

a. Open the compass so that the cover forms a straight edge with the base.

b. Position the eyepiece lens to the full upright position.

c. Place your thumb through the loop.

d. Establish a steady base with your third and fourth fingers.

e. Extend your index finger along the side of the compass.

f. Place the thumb of your other hand between the eyepiece and lens.

g. Extend the index finger along the remaining side of the compass.

h. Secure the remaining fingers around the fingers of the other hand.

i. Place your elbows firmly into your side.

Note: This will place the compass between your chin and your belt.

j. Turn your entire body toward the object.

k. Align the compass cover directly at the object.

l. Read the azimuth from beneath the fixed black index line.

Evaluation Preparation:

Setup: Provide the Soldier with the equipment and or materials described in the conditions statement.

Brief Soldier: Tell the Soldier what is expected of him by reviewing the task standards. Stress to the Soldier the importance of observing all cautions, warnings, and dangers to avoid injury to personnel and, if applicable, damage to equipment.

Performance Measures	GO	NO GO
1 Inspected the compass.	___	___
2 Determined direction.	___	___
3 Determined an azimuth using the compass-to-cheek method.	___	___
4 Determined an azimuth using the center-hold method.	___	___

Evaluation Guidance: Score the Soldier GO if all performance measures are passed. Score the Soldier NO-GO if any performance measure is failed. If the Soldier scores a NO-GO, show the Soldier what was done wrong and how to do it correctly.

References:
Required:
Related: TC 3-25.26

071-COM-1006

Navigate from One Point on the Ground to another Point while Dismounted

Conditions: You are a member of a squad or team in a field environment and have been directed to conduct movement to a designated point. You have a 1:50,000-scale topographic map of the area, a coordinate scale, a protractor, and a magnetic compass. Some iterations of this task should be performed in MOPP 4.

Standards: Navigate to the designated point using terrain association, dead reckoning, or a combination of both.

Special Condition: None

Special Standards: None

Safety Risk: Low

MOPP 4: Sometimes

Cue: None

Note: None

Performance Steps

1. Navigate using terrain association.
   a. Identify the start point and destination point on the map.
   b. Analyze the terrain between these two points for both movement and tactical purposes.
   c. Identify terrain features that can be recognized during movement, such as hilltops, roads, rivers, etc.
   d. Plan the best route, including checkpoints, if needed.
   e. Determine the map distances between identified checkpoints and the total distance to be traveled.
   f. Determine the actual ground distance by adding 20 percent to the map distance.
Note: Twenty percent is a general rule of thumb for cross country terrain - road movement and flat terrain do not require this 20 percent increase.
   g. Move to the designated end point (or intermediate point) using identified terrain features as aiming points or handrails.
Note: Handrails are linear features like roads or highways, railroads, power transmission lines, ridgelines, or streams that run roughly parallel to your direction of travel.

2. Navigate using dead reckoning.
Note: The use of steering marks is recommended when navigating by dead reckoning. A steering mark is a distant feature visible along one's route that is used as distant aiming point that one moves towards. Once reached another steering point is identified until a change of direction or the final destination is reached.
   a. Identify the start point and destination point on the map.
   b. Analyze the terrain between these two points for both movement and tactical purposes.
   c. Plan the best route, including checkpoints, if needed.
   d. Determine the grid azimuths between identified checkpoints (if any) and the final point.
   e. Convert the grid azimuth(s) taken from the map to a magnetic azimuth(s).
   f. Determine the map distances between identified checkpoints and the total distance to be traveled.
   g. Determine the direction of movement using the compass.

h. Move in the identified direction of travel or towards the identified steering mark.

i. Determine a new steering mark or confirm direction of travel as needed. *Note*: The direction of movement, when not using a steering mark, must be periodically confirmed.

3. Navigate using a combination of dead reckoning and terrain association.

a. Follow the procedures outlined for both techniques.

b. Use each technique to reinforce the accuracy of the other technique.

Evaluation Preparation:

Setup: Provide the Soldier with the equipment and or materials described in the conditions statement.

Brief Soldier: Tell the Soldier what is expected of him by reviewing the task standards. Stress to the Soldier the importance of observing all cautions, warnings, and dangers to avoid injury to personnel and, if applicable, damage to equipment.

Performance Measures	GO	NO GO
1 Navigated using terrain association.	_____	_____
2 Navigated using dead reckoning.	_____	_____
3 Navigated using a combination of dead reckoning and terrain association.	_____	_____

Evaluation Guidance: Score the Soldier GO if all performance measures are passed. Score the Soldier NO-GO if any performance measure is failed. If the Soldier scores a NO-GO, show the Soldier what was done wrong and how to do it correctly.

References:
Required:
Related: TC 3-25.26

071-COM-0501

Move as a Member of a Team

Conditions: You are a member of a dismounted team that is conducting tactical movement. You are not the team leader. You have your individual weapon and individual combat equipment. Some iterations of this task should be performed in MOPP 4.

Standards: Assume your position in the team's current formation, maintain proper distance between you and other team members, follow the team leader's example, and maintain security of your sector.

Special Condition: None

Special Standards: None

Safety Risk: Medium

MOPP 4: Sometimes

Cue: None

*Note:*The standard team is composed of four personnel - team leader (TL), automatic rifleman (AR), grenadier (G), and rifleman (R). The team leader designates positions based on the mission variables.

Performance Steps
1. Assume your position in the team's current formation.
Note: Specific positions vary based on the type of movement formation selected by the team leader.
   a. Assume your position within the team wedge formation (Figure 071-COM-0501-1).

Note: This is the basic team formation. It is easy to control, is flexible, allows immediate fires in all directions, and offers all-round local security.

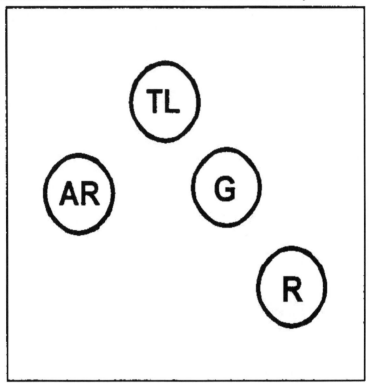

Figure 071-COM-0501-1. Wedge Formations

b. Assume your position within the team file formation (Figure 071-COM-0501-2).

Note: The file is used when employing the wedge is impractical. This formation is most often used in severely restrictive terrain, like inside a building; dense vegetation; limited visibility; and so forth. The distance between Soldiers changes due to constraints of the situation, particularly when in urban operations.

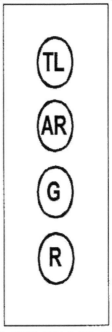

Figure 071-COM-0501-2. File Formation.

2. Maintain proper distance between you and other team members.
Note: The normal distance between Soldiers is 10 meters. When enemy contact is possible, the distance between teams should be about 50 meters. In open terrain such as desert, the interval may increase. The distance between individuals is determined by how much control the team leader can still exercise over his team members.

3. Maintain visual contact with your team leader.
Note: It is essential for all team members to maintain visual contact with the team leader.

4. Follow the team leader's example.
Note: When the team leader moves left, you move to the left. When the team leader gets down, you get down.

5. Adjust your position within the team as designated by the team leader.

6. Maintain security of your sector (i.e. to the flanks, front or rear of the team).

Evaluation Preparation:

Setup: Provide the Soldier with the equipment and or materials described in the conditions statement.

Brief Soldier: Tell the Soldier what is expected of him by reviewing the task standards. Stress to the Soldier the importance of observing all cautions, warnings, and dangers to avoid injury to personnel and, if applicable, damage to equipment.

Performance Measures	GO	NO GO
1 Assumed position in the team's current formation	____	____
2 Maintained proper distance from other team members.	____	____
3 Maintained visual contact with the team leader.	____	____
4 Followed the team leader's example.	____	____
5 Changed position within the team as designated by the team leader.	____	____
6 Maintained security of assigned sector.	____	____

Evaluation Guidance: Score the Soldier GO if all performance measures are passed. Score the Soldier NO-GO if any performance measure is failed. If the Soldier scores a NO-GO, show the Soldier what was done wrong and how to do it correctly.

References:
Required:
Related: TC 3-21.75

071-COM-0502

Move Under Direct Fire

Conditions: You are a member of a team conducting movement to contact and are under fire from an enemy position that is 250 to 300 meters away from your position. You have an individual weapon, individual combat equipment, and a current firing position that provides cover from the enemy's direct fire. Some interation of this task should be performed in MOPP 4.

Standards: Move within 100 meters of the enemy position using the appropriate movement techniques based on the situation and terrain.

Special Condition: None

Safety Risk: Medium

Special Equipment:

MOPP 4: Sometimes

Cue: None

Note: While this task may be performed by an individual Soldier, it is best performed as a member of a team or as part of a two-man buddy team.

Performance Steps

1. Select an individual movement route that adheres to the instructions provided by your team leader.
Note: When part of a team your movement route and general firing positions may be determined by your team leader. When moving as part of a team you must be prepared to follow your team leader's example.
 a. Search the terrain to your front for good firing positions.
Note: Large trees, rocks, stumps, fallen timber, rubble, vehicle hulls, man-made structures, and folds or creases on the ground may provide both cover and concealment and can be used as fighting positions.
 b. Select the best route to the positions.
Note: A gully, ravine, ditch, or wall at a slight angle to your direction of travel may provide cover and concealment when using the low or high crawl movement techniques. Hedge rows or a line of thick vegetation may provide concealment only when using the low or high crawl technique.
  (1) Pick a route that minimize your exposure to enemy fire.
  (2) Ensure route does not cross in front of other team members.
2. Communicate your movement intent to your buddy and team leader, as appropriate, using hand and arm signals.
3. Suppress the enemy as required.
Note: Do not expose yourself to fire unless the enemy is suppressed. Suppression of the enemy may be accomplished by another element, a buddy, or by yourself. With the enemy suppressed you can select an individual movement route or initiate movement.

4. Conduct movement using the appropriate technique(s) to reach each position.

   a. Move using the high crawl technique (figure 071-COM-1502-1).

Note: The high crawl lets you move faster than the low crawl and still gives you a low silhouette. Use this crawl when there is good cover and concealment but enemy fire prevents you from getting up.

Figure 071-COM-1502-1. High Crawl.

(1) Keep your body off of the ground.

(2) Rest your weight on your forearms and lower legs.

(3) Cradle your weapon in your arms.

(4) Keep the muzzle of the weapon off the ground.

(5) Keep your knees well behind your buttocks so it stays low.

(6) Move forward by alternately advancing your right elbow and left knee, and left elbow and right knee.

   b. Move using the low crawl technique (figure 071-COM-1502-2).

Note: The low crawl gives you the lowest silhouette. It is used to cross places where the cover and/or concealment are very low and enemy fire or observation prevents you from getting up.

Figure 071-COM-1502-2. Low Crawl.

(1) Keep your body as flat as possible to the ground.

(2) Grasp the sling of the weapon at the upper sling swivel with your right hand.

(3) Let the hand guard rest on your forearm.

(4) Keep the muzzle of the weapon off the ground.

(5) Move forward.

(a) Push both arms forward while pulling your right leg forward.

(b) Pull on the ground with both arms while pushing with your right leg.

(c) Repeat steps (a) and (b) until you reach your next position.

c. Moved using the rush technique (figure 071-COM-1502-3).

Note: The rush is the fastest way to move from one position to another. Use when you must cross an open area and time is critical.

Figure 071-COM-1502-3. Rush.

(1) Raise your head.

(2) Select your next position.

(3) Lower your head.

(4) Draw your arms into your body.

(5) Pull your right leg forward.

(6) Raise your body.

(7) Get up quickly.

(8) Run for 3-5 seconds to your next position.

(9) Plant both feet just before hitting the ground.

(10) Fall forward.

(a) Drop to your knees.

(b) Slide your right hand down to the heel of the butt of your weapon.

(c) Break your fall with the butt of your weapon.

d. Continue using movement techniques until you reach your final firing position.

5. Occupy your identified firing position within 100 meters of the enemy position.

a. Assume a firing position.

b. Engage enemy with your individual weapon.

Evaluation Preparation:

Setup: Provide the Soldier with the equipment and/or materials described in the conditions statement.

Brief Soldier: Tell the Soldier what is expected of him by reviewing the task standards. Stress to the Soldier the importance of observing all cautions, warnings, and dangers to avoid injury to personnel and, if applicable, damage to equipment.

	Performance Measures	GO	NO GO
1	Selected an individual movement route that adhered to the instructions provided by your team leader.	____	____
2	Communicated movement intent to buddy and team leader, as appropriate, using hand and arm signals.	____	____
3	Suppressed the enemy as required.	____	____
4	Conducted movement using the appropriate technique(s) to reach each position.	____	____
5	Occupied your identified firing position within 100 meters of the enemy position.	____	____

Performance Measures	GO	NO GO

Evaluation Guidance: Score the Soldier GO if all performance measures are passed. Score the Soldier NO-GO if any performance measure is failed. If the Soldier scores a NO-GO, show the Soldier what was done wrong and how to do it correctly.

References:
Required:
Related: TC 3-21.75

071-COM-0510

React to Indirect Fire while Dismounted

Conditions: You are a member of a squad or team conducting a dismounted patrol and you hear indirect fire exploding or passing Over-head. You have your individual weapon and equipment. Some iterations of this task should be performed in MOPP 4.

Standards: React to indirect fire while moving as a member of a squad or team.

Special Condition: None

Safety Risk: Medium

MOPP 4: Sometimes

Cue: None

Performance Steps

    1. Shout "Incoming!" in a loud, recognizable voice.
    2. Drop to the ground.
    3. Follow commands and actions of your leader.
Note: Normally, if moving, the leader will tell you to run out of the impact area in a certain direction or will tell you to follow him. If you cannot see or hear your leader you should follow other team members.
    4. Seek the nearest appropriate cover.
    5. Avoid the impact area if not already in it.
    6. Run in a direction away from the incoming fire.
    7. Assess your situation.
    8. Report your situation to your leader.
    9. Continue the mission.

Evaluation Preparation: *Setup:* Provide the Soldier with the equipment and/or materials described in the conditions statement.
Brief Soldier: Explain what is expected from the Soldier by reviewing the task standards. Stress the importance of observing all cautions, warnings, and dangers to avoid injury to personnel and, if applicable, damage to equipment.

Performance Measures	GO	NO GO
1 Shouted "Incoming!" in a loud, easily recognizable voice.	____	____
2 Drop to the ground.	____	____
3 Followed the commands and actions of your leader.	____	____
4 Seeked the nearest appropriate cover.	____	____
5 Avoided the impact area if not already in it.	____	____
6 Ran in direction away from the incoming fire.	____	____
7 Assessed your situation.	____	____
8 Reported your situation to your leader.	____	____
9 Continued the mission.	____	____

Evaluation Guidance:. Refer to chapter 1, paragraph 1-9e, (1) and (2).

References: TC 3-21.75

1

071-COM-0513

Select Hasty Fighting Positions

Condition: You are a member of a dismounted squad or team occupying an area and have been directed to establish a temporary fighting position to cover a given sector of fire. You have an individual or crew-served weapon and your individual combat equipment.

Standard: Select and prepare a hasty fighting position that protects you from enemy observation and fire, and allows effective fires to be placed within sector of fire.

Special Condition: None

Safety Level: Low

Performance Steps

1. Identify a position that will provide the best cover and concealment.

Note: Cover, made of natural or man-made materials, gives protection from bullets, fragments of exploding rounds, flame, nuclear effects, biological and chemical agents, and enemy observation. Concealment is anything that hides personal, equipment and/or vehicles from enemy observation. Concealment does not protect you from enemy fire.

    a. Use natural, undisturbed cover and concealment, if available.

    b. Ensure man-made cover and concealment blends with surroundings.

2. Ensure the position allows effective weapon emplacement.

    a. Ensure proper sector of fires for appropriate weapon system.

    b. Ensure proper field of fires.

3. Prepare the fighting position.

    a. Avoid disclosing your position by careless or excessive clearing.

    b. Leave a thin, natural screen of vegetation to hide your position.

    c. Cut off lower branches of large, scattered trees, in sparsely wooded areas.

    d. Clear underbrush only where it blocks your view.

    e. Remove cut brush, limbs, and weeds so the enemy will not spot them.

    f. Cover cuts on trees and bushes forward of your position with mud, dirt, or snow.

    g. Leave no trails as clues for the enemy.

4. Maintain camouflage.

Note: Camouflage is anything you use to keep yourself, your equipment, and your position from being identified.

    a. Prevent attention by controlling movement and activities.

    b. Avoid putting anything where the enemy expects to find it.

    c. Break up outlines and shadows.

d. Conceal shining objects.

e. Break up familiar shapes to make them blend in with their surroundings.

f. Camouflage yourself and your equipment to blend with the surroundings.

g. Ensure proper dispersion.

h. Study the terrain and vegetation of the area in which you are operating.

i. Use camouflage material that best blends with the area.

Evaluation Guidance: Score the Soldier GO if all performance measures are passed. Score the Soldier NO-GO if any performance measure is failed. If the Soldier scores a NO-GO, show the Soldier what was done wrong and how to do it correctly.

Evaluation Preparation: SETUP: Provide the Soldier with the equipment and/or materials described in the conditions statement.

BRIEF THE SOLDIER: Tell the Soldier what is expected by reviewing the task standards. Stress to the Soldier the importance of observing all cautions, warnings, and dangers to avoid injury to personnel and, if applicable, damage to equipment.

Performance Measures		GO	NO GO
1	Identified a position that provided the best cover and concealment.	____	____
2	Ensured the position allowed effective weapon emplacement.	____	____
3	Prepared the fighting position.	____	____
4	Maintained camouflage.	____	____

Environment: Environmental protection is not just the law but the right thing to do. It is a continual process and starts with deliberate planning. Always be alert to ways to protect our environment during training and missions. In doing so, you will contribute to the sustainment of our training resources while protecting people and the environment from harmful effects. Refer to FM 3-34.5 Environmental Considerations and GTA 05-08-002 ENVIRONMENTAL-RELATED RISK ASSESSMENT. Environmental protection is not just the law but the right thing to do. It is a continual process and starts with deliberate planning. Units will assess environmental risk using the checklist in TC 3-34.489 and assessment matrixes in FM 3-34.5, Appendix D. Always be alert to ways to protect our environment during training and missions. In doing so, you will contribute to the sustainment of our training resources while protecting people and the environment from harmful effects.

Safety: In a training environment, leaders must perform a risk assessment in accordance with ATP 5-19, Risk Management. Leaders will complete a DD Form 2977 DELIBERATE RISK ASSESSMENT WORKSHEET during the planning and completion of each task and sub-task by assessing mission, enemy, terrain and weather, troops and support available-time available and civil considerations, (METT-TC). Note: During MOPP training, leaders must ensure personnel are monitored for potential heat injury. Local policies and procedures must be followed during times of increased heat category in order to avoid heat related injury. Consider the MOPP work/rest cycles and water replacement guidelines IAW FM 3-11.4, Multiservice Tactics, Techniques, and Procedures for Nuclear, Biological, and Chemical (NBC) Protection, FM 3-11.5, Multiservice Tactics, Techniques, and Procedures for Chemical, Biological, Radiological, and Nuclear Decontamination.

References:
Required:
Related: TC 3-21.75

Subject Area 3: Communicate

113-COM-2070

Operate SINCGARS Single-Channel (SC)

Conditions: Given an operational SINCGARS, Army standard Data Transfer Device (DTD), distant station, TM 11-5820-890-10-8, TM 11-5820-890-10-3, ACP 125 US Suppl-1, and unit SOI or ANCD w/SOI data loaded.

Standards: Operate a SINCGARS in SC mode that results in a secure communications check with a distant station.

Special Condition: None

Safety Risk: Low

Performance Steps
    1. Perform starting procedures.
    2. Load the traffic encryption key (TEK).
    3. Enter the net.

      a. Use the correct procedures.

      b. Conduct a secure communications check

    4. Prepare control monitor (CM) for operation.

    5. Change the radio functions using the control monitor.

    6. Perform stopping procedures.

Evaluation Preparation:

Performance Measures	GO	NO GO
1 Performed starting procedures.	____	____
2 Load traffic encryption key (TEK).	____	____
3 Entered net.	____	____
4 Prepared control monitor (CM) for operation.	____	____
5 Changed the radio functions using control monitor.	____	____
6 Performing stopping procedures.	____	____

Evaluation Guidance: Score the soldier GO if all steps are passed. Score the soldier NO-GO if any step is failed. If the soldier fails any step, show what was done wrong and how to do it correctly. Have the soldier practice until the task can be performed correctly.

References
Required: ACP 125 US SUPP-1, TM 11-5820-890-10-1, TM 11-5820-890-10-3, TM 11-5820-890-10-8, UNIT SOI Unit/Unit's Signal Operation Instructions (SOI)
Related

113-COM-1022

Perform Voice Communications

Conditions: Given: 1. One operational radio set for each net member, warmed up and set to the net frequency. 2. A call sign information card (5 x 8) consisting of: Net member duty position (S-1, S-2), net call sign (letter-number-letter), suffix list (Net Control Station [NCS] - 46, S-1 - 39, S-2 - 13), and a message to be transmitted. 3. Situation: The net is considered to be secure and authentication is not required. Note: This task may have as many net members as there is equipment available. Each net member must have a different suffix and message to transmit.

Standards: Enter a radio net, send a message, and leave a radio net using the proper call signs, call sign sequence, prowords, and phonetic alphabet and numerals with 100 percent accuracy.

Special Condition: None

Special Standards: None

Safety Risk: Low

Cue: None

Note: None

Performance Steps

1. Enter the net.
   a. Determine the abbreviated call sign and answering sequence for your duty position.
   b. Respond to the NCS issuing a net call.
   c. Answer in alphanumeric sequence.
Note: At this time, the NCS acknowledges and the net is open.
2. Send a message.
   a. Listen to make sure the net is clear. Do not interrupt any ongoing communications.
   b. Call the NCS and tell the operator the priority of the message you have for his or her station.
   c. Receive a response from the NCS that he or she is ready to receive.
   d. Send your message using the correct prowords and pronunciation of letters and numbers.
   e. Get a receipt for the message.
3. Leave the net in alphanumeric sequence.

Note: The NCS acknowledges and the net is closed. Note: The following call signs are used in this task as an example: Net call sign - E3E, NCS - E46, S-1 - E39, S-2 - E13.

    a. Answer in alphanumeric sequence.

    b. You receive a call from the NCS who issues a close down order.

Evaluation Preparation: *Setup*: Position operational radio sets in different rooms or tents or at least 70 feet apart outside. Obtain call signs, suffixes, and a radio frequency through the normal command chain. Select a message 15-25 words in length, containing some number groups such as map coordinates and times. Print the call signs for the sender and the receiver, along with the message to be sent, on 5 x 8 cards. Perform a communications check to ensure operation of the radios. Have an assistant who is proficient in radio operation man the NCS. Provide the assistant with the call signs. If the soldier has not demonstrated sufficient progress to complete the task within 5 minutes, give him or her a NO-GO. This time limit is an administrative requirement, not a doctrinal one; so if the soldier has almost completed the task correctly, you may decide to allow him or her to finish.

Brief Soldier: Give the soldier to be tested the card containing the message and call signs. Tell him or her the radio is ready for operation, the net is considered to be secure and authentication is not required, and to send the message to the NCS and get a receipt. Tell the soldier, if sufficient progress in completing the task within 5 minutes has not been demonstrated, he or she will receive a NO-GO for the task.

Performance Measures	**GO**	**NO GO**
1 Entered the net in alphanumeric sequence.	____	____
2 Sent a message of 15 to 25 words using the correct prowords and phonetic alphabet and numerals.	____	____
3 Left the net in alphanumeric sequence.	____	____

Evaluation Guidance: Score the soldier GO if all performance measures are passed. Score the soldier NO-GO if any performance measure is failed. If the soldier scores NO-GO, show the soldier what was done wrong and how to do it correctly.

References:
Required:
Related: ACP 125 (F); ACP 131 (E);
TB 9-2320-280-35-2

171-COM-4079

Send a Situation Report (SITREP)

Conditions: You are an element leader with an operation order (OPORD) or fragmentary order (FRAGO), map, overlay or sketch map with graphic control measures, and an operational vehicle. You may be digitally equipped. Your current situation requires you to send a SITREP. Some iterations of this task should be performed in MOPP 4.

Standards: Prepare a SITREP in standard format and send to the next higher element. Maintain situational awareness (SA).

Special Condition: None

Special Standards: None

Special Equipment:

Cue: None

*Note:*The operational environment must be considered at all times during this task. All Army elements must be prepared to enter any environment and perform their missions while simultaneously dealing with a wide range of unexpected threats and other influences. Units must be ready to counter these threats and influences and, at the same time, be prepared to deal with various third-party actors, such as international humanitarian relief agencies, news media, refugees, and civilians on the battlefield. These groups may or may not be hostile to us, but they can potentially affect the unit's ability to accomplish its mission.

Note: Units equipped with digital communication systems will use these systems to maximize information management, maintain SA, and minimize electronic signature.

Performance Steps
   1. Prepare a SITREP in standard format.

Note: The SITREP is used to report any change since the last report, to request resupply, and to report the current location of the element; only lines or parts of lines that contain new information will be sent. It may require additional follow-up reports.

Note: Timely and accurate reporting of friendly elements locations, obstacles and contacts are essential to maintaining SA and the reduction of potential fratricide incidents.

    a. Line 1: Date and Time Group (DTG)-Report date and time the report is being submitted.

Note: Date is the date that the report is being submitted. Time is the local time or zulu time that the report is being initiated.

    b. Line 2: Unit-Identify the unit making the report.

    c. Line 3: From-Report the time that the operational situation started or will start.

    d. Line 4: Until-Report the time that the operational situation ends or will end.

    e. Line 5: Map-Give a minimum six digit grid of the squad or team current location.

    f. Line 6: Enemy-Report enemy activity.

        (1) Nationality.

        (2) Location.

        (3) Mission.

        (4) Time of Sighting.

    g. Line 7: Nonhostile-Report nonhostile activity.

    h. Line 8: Own-Report activities of own forces.

        (1) Changes in location of units and/or formations.

        (2) Activities of forces not attached to originating unit.

  2. Send the SITREP to the next higher element.

  3. Maintain SA.

Evaluation Preparation:

Setup: Provide the Soldier with the equipment and or materials described in the conditions statement.

Brief Soldier: Tell the Soldier what is expected of him by reviewing the task standards. Stress to the Soldier the importance of observing all cautions, warnings, and dangers to avoid injury to personnel and, if applicable, damage to equipment.

Performance Measures	GO	NO GO
1 Prepared the SITREP in standard format.	_____	_____

Performance Measures	GO	NO GO
2 Sent the SITREP.	___	___
3 Maintained SA.	___	___

Evaluation Guidance: Score the Soldier GO if all performance measures are passed. Score the Soldier NO-GO if any performance measure is failed. If the Soldier scores a NO-GO, show the Soldier what was done wrong and how to do it correctly.

References:
Required: FM 6-99
Related:

171-COM-4080

Send a Spot Report (SPOTREP)

Conditions: You are an element leader with in an operation environment. You may be digitally equipped. Your current situation requires you to send a Spot Report (SPOTREP). Some iterations of this task should be performed in MOPP 4.

Standards: Prepare a Spot Report (SPOTREP) in standard format and send to the next higher element.

Special Condition: None

Safety Risk: Low

MOPP 4: Sometimes

Cue: None

Note: The SPOTREP is used to report timely intelligence or status regarding events that could have an immediate and significant effect on current and future operations. This is the initial means for reporting troops in contact and event information. Several lines of the SPOTREP provide sub-categories that structure reported data. Some lines may be omitted in an emergency. For example, the SPOTREP could provide only the reporting unit, event DTG, location, and activity. The format of a SPOTREP may also change based on unit's standing operating procedures (SOP).

If equipped with Force XXI Battle Command Brigade-and-Below (FBCB2), the FBCB2 operator must update observed enemy force locations, neutral organizations, civilians and other battlefield hazards.

Performance Steps

1. Prepare SPOTREP.
    a. LINE 1 – date time group (DTG) of report submission.
    b. LINE 2 – reporting unit (Unit Making Report).
Note: After the unit designation, the method of observation must be indicated: unaided, binoculars, infrared, thermal, night vision device (NVD), unmanned aircraft system (UAS), or other. Follow with narrative if needed.
    c. LINE 3 – size of detected element.
       (1) Persons: Military, Civilian.
       (2) Vehicles: Military, Civilian.
       (3) Equipment: Military, Civilian
    d. LINE 4 - activity of detected element at DTG of report.
Note: The activity type or types must be indicated and an amplifying sub-type if Applicable. If necessary add a narrative to clarify, describe, or explain the type of activity.
       (1) Attacking (direction from).
         (a) Air defense artillery (ADA) (engaging).
         (b) Aircraft (engaging) (rotary wing [RW], fixed wing [FW]).
         (c) Ambush (IED [exploded], IED [unexploded], sniper, anti-armor, other).
         (d) Indirect fire (point of impact, point of origin).
         (e) Chemical, biological, radiological or nuclear (CBRN).
       (2) Defending (direction from).
       (3) Moving (direction from).
       (4) Stationary.
       (5) Cache.
       (6) Civilian (criminal acts, unrest, infrastructure damage).
       (7) Personnel recovery (isolating event, observed signal).
       (8) Other (give name and description).
    e. LINE 5 - location (universal transverse mercator (UTM) or grid coordinate with military grid reference system (MGRS) grid zone designator of detected element activity or event observed).
    f. LINE 6 - unit (detected element unit, organization, or facility).
Note: The type of unit, organization, or facility detected should be identified. If it cannot be clearly identified is should be described in as much detail as possible to include; the type uniform, vehicle markings, and other identifying information.
       (1) Conventional.
       (2) Irregular.
       (3) Coalition.
       (4) Host nation.
       (5) Nongovernmental organization (NGO).
       (6) Civilian.

(7) Facility.

g. LINE 7 – time (DTG of observation).

h. LINE 8 – equipment (equipment of element observed).

Note: The equipment type or types, and amplifying sub-type should be identified, if applicable. A narrative can be added if necessary to clarify, describe, or explain the type of equipment. The nomenclature, type, and quantity of all equipment observed should be provided, if known. If equipment cannot be clearly identified it should be describe in as much detail as possible

(1) ADA (missile (man-portable air defense system [MANPADS]), missile (other), gun).

(2) Arty (gun (self-propelled [SP]), gun (towed), missile or rocket, mortar).

(3) Armored track vehicle (tank, armored personnel carrier [APC], command and control [C2], engineer, transport, other).

(4) Armored wheel vehicle (gun, APC, C2, engineer, transport, other).

(5) Wheel vehicle (gun, C2, engineer, transport, other).

(6) INF weapon (WPN) (anti-armor missile, anti-armor gun, rocket-propelled grenade [RPG], heavy [HVY] machine gun [MG], grenade launcher [GL], small arms, other).

(7) Aircraft (RW (attack helicopter [AH]), RW (utility helicopter [UH]), RW (observation helicopter), FW (atk), FW (trans), UAS, other).

(8) Mine or IED (buried, surface, vehicle-borne improvised explosive device [VBIED], person-borne improvised explosive device [PBIED], other).

(9) CBRN.

(10) Supplies (class III, class V, other).

(11) Civilian.

(12) Other.

i. LINE 9 – assessment (apparent reason for or purpose of the activity observed, and apparent threats to or opportunities for friendly forces).

j. LINE 10 –narrative (free text for clarifying report).

Note: The narrative should describe the actions taken related to the detected activity: attack, withdraw, continue to observe, or other. When feasible, the narrative should also state potential for subsequent reports such as air support request, battle damage assessment (BDA) report, call for fire, casualty report, explosive ordinance disposal (EOD) support, medical evacuation (MEDEVAC) or other reports.

k. LINE 11 – authentication (report authentication) per SOP.

2. Send SPOTREP to next higher element.

Note: The unit SOP may have additional guidance on who receives the SPOTREP.

Evaluation Preparation: *Setup*: Provide the Soldier with the equipment and or materials described in the conditions statement.

Brief Soldier: Tell the Soldier what is expected of him by reviewing the task standards. Stress to the Soldier the importance of observing all cautions, warnings, and dangers to avoid injury to personnel and, if applicable, damage to equipment.

Performance Measures	GO	NO GO
1 Prepared the SPOTREP.	_____	_____
2 Sent the SPOTREP to higher headquarters.	_____	_____

Evaluation Guidance: Score the Soldier GO if all performance measures are passed. Score the Soldier NO-GO if any performance measure is failed. If the Soldier scores a NO-GO, show the Soldier what was done wrong and how to do it correctly.

References:
Required: FM 6-99
Related:

071-COM-0608

Use Visual Signaling Techniques

Conditions: You are a member of a mounted or dismounted platoon in a field environment and must use visual signals to communicate. Some iterations of this task should be performed in MOPP 4.

Standards: Communicate with other Soldiers and vehicle crews using visual signaling techniques.

Special Condition: None

Safety Risk: Low

MOPP 4: Sometimes

Cue: None

Note: Visual signals are any means of communication that require sight and can be used to transmit planned messages rapidly over short distances. This includes the devices and means used for the recognition and identification of friendly forces. The most common types of visual signals are arm-and-hand, flag, pyrotechnic, and ground-to-air. However, Soldiers are not limited to the types of signals discussed and may use what is available. Chemical light sticks, flashlights, and other items can be used, provided their use is standardized within a unit and understood by Soldiers and units working in the area. The only limit is the Soldier's initiative and imagination. Visual signals have certain limitations: (1) range and reliability of visual communications are significantly reduced during poor visibility and when terrain restricts observation; (2) may be misunderstood; and (3) vulnerable to enemy interception and may be used for deception. Leaders of mounted units use arm-and-hand signals to control individual vehicles and platoon movement. When distances between vehicles increase, flags can be used as an extension of the arms to give the signals.

Performance Steps

1. Use visual signals for combat formations

    a. Disperse (Figure 071-326-0608-2)
      (1) Extend the arm horizontally.
      (2) Wave the arm and hand to the front, left, right, and rear
      (3) Point toward the direction of each movement.

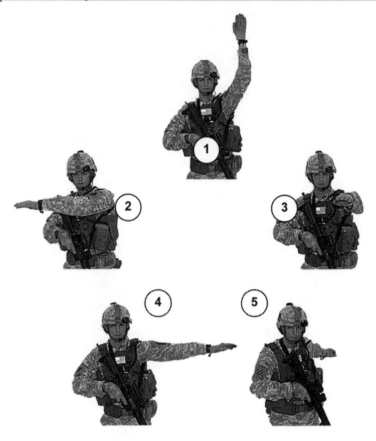

Figure 071-COM-0608-1. Disperse.

b. Assemble or Rally (Figure 071-326-0608-2).
Note: The assemble and rally signal is normally followed by pointing to the assembly or rally site.
    (1) Raise arm vertically overhead.
    (2) Turn palm to the front.
    (3) Wave in large horizontal circles.

**Figure 071-COM-0608-2.
Assemble or Rally.**

c. Join me, Follow me, or Come forward (Figure 071-COM-0608-3).
   (1) Point toward person(s) or unit.
   (2) Beckon by holding the arm horizontally to the front with palm up.
   (3) Motion toward the body.

Figure 071-COM-0608-3.
Join me, Follow me, or come forward.

    d. Increase speed, Double time, or Rush (Figure 071-COM-0608-4).
      (1) Raise the fist to the shoulder.
      (2) Thrust the fist upward to the full extent of the arm and back to shoulder level.
      (3) Continue rapidly several times.

Figure 071-COM-0608-4.
Increase speed, Double time, or Rush.

    e. Quick time (Figure 071-COM-0608-5).

Note: This is the same signal as SLOW DOWN when directing vehicles. The difference in meaning must be understood from the context in which they are used.

     (1) Extend the arm horizontally sideward.

     (2) Turn palm to the front.

     (3) Wave the arm slightly downward several times, keeping the arm straight.

     (4) Keep arm at shoulder level.

Figure 071-COM-0608-5.
Quick Time.

   f. Enemy in sight (Figure 071-COM-0608-6).

     (1) Hold the rifle in the ready position at shoulder level.

     (2) Point the rifle in the direction of the enemy.

**Figure 071-COM-0608-6.
Enemy in sight.**

g. Wedge (Figure 071-COM-0608-7).
　(1) Extend the arms downward to the side.
　(2) Turn the palms to the front.
　(3) Place your arms at a 45-degree angle below horizontal.

**Figure 071-COM-0608-7.
Wedge.**

h. Vee (Figure 071-COM-0608-8).
   (1) Raise the arms.
   (2) Extend the arms 45-degrees above the horizontal.

**Figure 071-COM-0608-8.
Vee.**

i. Line (Figure 071-COM-0608-9).
   (1) Extend the arms.
   (2) Turn palms downward parallel to the ground.

**Figure 071-COM-0608- 9.
Line.**

j. Coil (Figure 071-COM-0608-10
   (1) Raise one arm above the head.
   (2) Rotate it in a small circle.

Figure 071-COM-0608-10.
Coil.

k. Staggered Column (Figures 11).
    (1) Extend the arms so that upper arms are parallel to the ground.
    (2) Make sure the forearms are perpendicular.
    (3) Raise the arms so they are fully extended above the head.

Figure 071-COM-0608-11.
Staggered Column.

2. Use visual signals for battle drills.

Note: Drills are a rapid, reflexive response executed by a small unit. These signals are used to initiate drills.

    a. Contact left (Figure 071-COM-0608-12).

        (1) Extend the left arm parallel to the ground.

        (2) Bend the arm until the forearm is perpendicular.

        (3) Repeat.

Figure 071-COM-0608-12.
Contact left.

b. Contact right (Figure 071-COM-0608-13).
   (1) Extend the both arms parallel to the ground.
   (2) Bend the arm until the forearm is perpendicular.
   (3) Repeat.

Figure 071-COM-0608-13.
Contact right.

c. Action left (Figure 071-COM-0608-14).
   (1) Extend the both arms parallel to the ground.
   (2) Raise the right arm until it is overhead.
   (3) Repeat.

Figure 071-COM-0608-14.
Action left.

   d. Action right (Figure 071-COM-0608-15).
     (1) Extend the both arms parallel to the ground
     (2) Raise the left arm until it is overhead.
     (3) Repeat.

Figure 071-COM-0608-15.
Action right.

  e. Air attack (Figure 071-COM-0608-16).
     (1) Bend the arms with forearms at a 45-degree angle.
     (2) The forearms should cross.
     (3) Repeat.

Figure 071-COM-0608-16.
Air Attack

 f. Nuclear, Biological, Chemical attack (Figure 071-COM-0608-17).
  (1) Extend the arms and fists.
  (2) Bend the arms to the shoulders.
  (3) Repeat.

Figure 071-COM-0608-17
Nuclear, Biological, Chemical attack.

3. Use visual signals for patrolling.
Note: Patrolling is conducted by many type units. Infantry units patrol in order to conduct combat operations. Other units patrol for reconnaissance and security. Successful patrols require clearly understood communication signals among members of a patrol.

    a. Map check (Figure 071-COM-0608-18).

       (1) Place one hand on top of the other.

       (2) Point at the palm of one hand with the index finger of the other hand.

Figure 071-COM-0608-18.
Map check.

b. Pace count (Figure 071-COM-0608-19).
  (1) Bend the knee so that the heel can be tapped on.
  (2) Tap the heel of the boot repeatedly with the open hand.

**Figure 071-COM-0608-19.
Pace Count.**

c. Head count (Figure 071-COM-0608-20).
   (1) Raise one arm behind the head.

(2) Tap the back of the helmet repeatedly with an open hand.

Figure 071-COM-0608-20.
Head count.

d. Danger area (Figure 071-COM-0608-21).
Note: This movement is the same as stop engine when directing a driver. The difference in meaning is understood from the context in which it is used.

(1) Raise the right hand up until it is level with the throat.

(2) Draw the right hand, palm down in a throat-cutting motion from left to right across the neck.

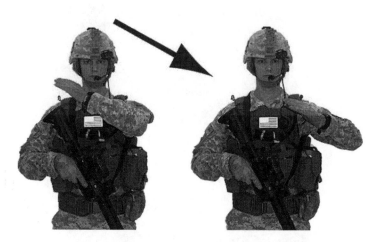

Figure 071-COM-0608-21.
Danger Area.

e. Freeze or halt (Figure 071-COM-0608-22).
　(1) Make a fist with the right hand.
　(1) Raise the fist to head level.

Figure 071-COM-0608-22.
Freeze or halt.

4. Use visual signals to control vehicle drivers.

Note: Flashlights or chemical lights are used at night to direct vehicles. Flashlights with blue filters and chemical lights will have less effect on a Soldier's night vision.

    a. Start engine or prepare to move.

       (1) Day: Simulate cranking of the engine by moving the arm, with the fist, in a circular motion at waist level (Figure 071-COM-0608-23).

Figure 071-COM-0608-23.
Start engine or prepare to move.

(2) Night: Move a light in a horizontal figure 8 in a vertical plane in front of the body (Figure 071-COM-0608-24).

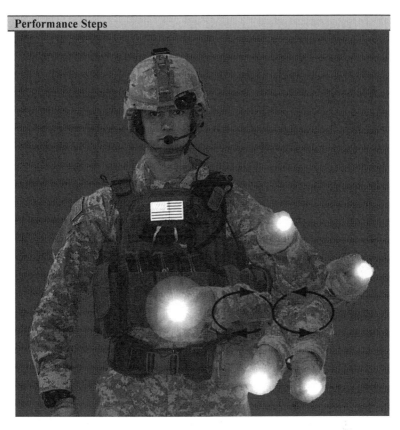

Figure 071-COM-0608-24.
Start engine, or prepare to move (night).

    b. Halt or stop.
       (1) Day (Figure 071-COM-0608-25).
         (a) Raise the hand upward to the full extent of the arm, with palm to the front.
         (b) Hold that position until the signal is understood.

Figure 071-COM-0608-25.
Halt or stop.

(2) Night (Figure 071-COM-0608-26).
(a) Move a light horizontally back and forth several times across the path of approaching traffic to stop vehicles.

(b) Use the same signal to stop engines.

**Figure 071-COM-0608-26.
Halt or Stop (night).**

b. Left turn.
(1) Day (Figure 071-COM-0608-27).
(a) Extend the right arm horizontally to the side.
(b) Turn palm toward vehicle with fingers extended in the direction of travel.

**Figure 071-COM-0608-27.
Left turn.**

(2) Night (Figure 071-COM-0608-28).
    (a) Bend the right arm at the elbow parallel to the ground.
    (b) Rotate a light to describe a 12 to 18 inch circle to the right.

Figure 071-COM-0608-28.
Left turn (night).

    d. Right Turn.
      (1) Day (Figure 071-COM-0608-29).
        (a) Extend the left arm horizontally to the side.
        (b) Turn palm toward vehicle with fingers extended in the direction
of travel.

**Figure 071-COM-0608-29.
Right turn.**

(1) Night (Figure 071-COM-0608-30).
    (a) Bend the right arm at the elbow parallel to the ground.
    (b) Rotate a light to describe a 12 to 18 inch circle to the left.

Figure 071-COM-0608-30.
Right turn (night).

    e. Move forward.

        (1). Day. (Figure 071-COM-0608-31).

           (a) Face the vehicle.

           (b) Raise the hands to shoulder level with palms facing the chest.

           (c) Move the hands and forearms backward and forward.

Figure 071-COM-0608-31.

Move forward.

(1). Night (Figure 071-COM-0608-32).
(a) Face the vehicle.
(b) Hold a light at shoulder level.
(c) Move the hands and forearms backward and forward.

Figure 071-COM-0608-32.
Move forward (night).

f. Move in reverse.
(1). Day (Figure 071-COM-0608-33).
(a) Face the vehicle.
(b) Raise the hands to shoulder level with palms facing the vehicle.
(c) Move the hands and forearms backward and forward.

Figure 071-COM-0608-33
Move in Reverse.

(2). Night (Figure 071-COM-0608-34).
  (a) Hold a light at shoulder level.
  (b) Blink it several times toward the vehicle.

Figure 071-COM-0608-34.
Move in reverse (night).

g. Stop engine
  (1). Day (Figure 071-COM-0608-35).

(a) Extend the arm parallel to the ground with hand open.

(b) Move the arm across the body in a throat-cutting motion.

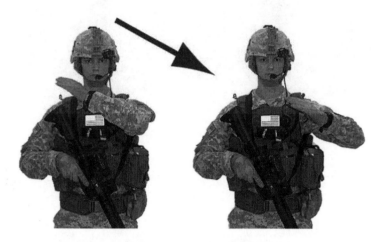

Figure 071-COM-0608-35.
Stop engine.

(2). Night (Figure 071-COM-0608-36).

(a) Extend the arm parallel to the ground with hand open.

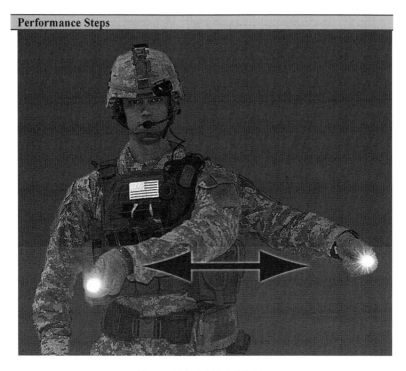

Figure 071-COM-0608-36.
Stop engine (night).

    b. Use the same signal to halt or stop vehicle.
(Asterisks indicates a leader performance step.)

Evaluation Preparation:

Setup: Provide the Soldier with the equipment and or materials described in the conditions statement.

Brief Soldier: Tell the Soldier what is expected of him by reviewing the task standards. Stress to the Soldier the importance of observing all cautions, warnings, and dangers to avoid injury to personnel and, if applicable, damage to equipment.

Performance Measures	GO	NO GO
1 Used visual signals for combat formations.	_____	_____
2 Used visual signals for battle drills.	_____	_____

Performance Measures	GO	NO GO
3 Used visual signals for patrolling.	_____	_____
4 Used visual signals to control vehicle drivers.	_____	_____

Evaluation Guidance: Score the Soldier GO if all performance measures are passed. Score the Soldier NO-GO if any performance measure is failed. If the Soldier scores a NO-GO, show him what was done wrong and how to do it correctly.

References
Required: FM 21-60
Related:

Subject Area 4: Survive

031-COM-1036

Maintain Your Assigned Protective Mask

WARNING
READ AND ADHERE TO ALL SAFETY NOTES IN YOUR MASK'S OPERATOR'S TM PRIOR TO BEGINNING MASK MAINTENANCE.

Conditions: You have used your assigned protective mask or must conduct a scheduled mask inspection. You have your assigned protective mask (with authorized accessories and components), cleaning materials in accordance with (IAW) the applicable operator technical manual (TM), a preventive maintenance checks and services (PMCS) Department of the Army (DA) Form 5988-E Equipment Inspection Maintenance Worksheet (EGA) or DA Form 2404 Equipment Inspection and Maintenance Worksheet IAW DA Pamphlet (PAM) 750-8, mask replacement parts and a new filter. This task cannot be performed in mission-oriented protective posture (MOPP) 4.

Standards: Maintain your assigned protective mask IAW the operator's TM by:
1) Performing operator's PMCS.
2) Cleaning your mask.

Special Condition: None

Special Standards: None

Safety Risk: Low

Cue: None

Note: None

Performance Steps

1. Inspect your protective mask, carrier, hood, and accessories according to the PMCS tables located in the mask operator TM.
   a. Identify deficiencies and shortcomings.
   b. Correct operator level deficiencies.
2. Clean and dry the mask, hood, and authorized accessories and components IAW the mask operator TM.
3. Record uncorrected deficiencies on DA Form 2404 Equipment Inspection and Maintenance Work Sheet IAW DA Pam 750-8.
4. Provide the completed DA Form 2404 Equipment Inspection and Maintenance Work Sheet to your supervisor for his/her review and guidance.
5. Perform all maintenance without damaging your protective mask.

Evaluation Preparation: *Setup*: A good time to evaluate this task is during normal care and cleaning of the mask. Place the required equipment on a field table or another suitable surface. Simulate defects in the mask by removing components from the mask or using a defective mask not issued to the Soldier. During training and evaluation sessions, use an old set of filters or canister several times to avoid expending new ones each time. If the Soldier has not made adequate progress towards completing the task within 30 minutes, stop him and give him a NO-GO. This time standard is administrative.

Brief Soldier: Tell the Soldier there is no time standard for this task on the job, but for testing purposes he must perform the task within 30 minutes. Tell him to perform operator level PMCS on the mask, clean his assigned protective mask, and replace the mask filter. Tell the Soldier that completing a DA Form 2404 Equipment Inspection and Maintenance Work Sheet IAW DA PAM 750-8 is not part of the task.

Performance Measures	GO	NO GO
1 Inspected protective mask, carrier, hood, and accessories according to the PMCS tables located in mask operator TM.	____	____
a. Identify deficiences and shortcomings.		

Performance Measures	GO	NO GO
b. Corrected operator level deficiences.		
2 Cleaned and dried the mask, hood, and authorized accessories and components IAW mask operator TM.	_____	_____
3 Recorded uncorrected deficiencies on a DA Form 2404 Equipment Inspection and Maintenance Work Sheet IAW DA Pam 750-8.	_____	_____
4 Provided the completed DA Form 2404 Equipment Inspection and Maintenance Work Sheet to his/her supervisor for review and guidance.	_____	_____
5 Performed all maintenance without damaging the protective mask.	_____	_____

Evaluation Guidance: Score the Soldier GO if all performance measures are passed (P). Score the Soldier NO-GO if any performance measure is failed (F). If the Soldier fails any performance measure, show him how to do it correctly.

References:
Required: PAM 750-8, TM 3-4240-312-12&P, TM 3-4240-342-10, TM 3-4240-346-10, TM 3-4240-348-10, TM 3-4240-542-13&P
Related:

031-COM-1035

Protect Yourself from Chemical And Biological (CB) Contamination Using Your Assigned Protective Mask

Conditions: You are given your assigned protective mask, hood, carrier, a canteen with an M1 canteen cap or water canteen cap, and M8 detector paper. You find yourself in one of the following situations: 1) you hear or see a CB agent and/or unknown toxic industrial chemical attack/spill, 2) you realize, through other means, that you are under a CB agent and/or toxic industrial chemical attack,

3) you are ordered to mask, 4) you must enter a contaminated area, and 5) after having donned your protective mask, you need to drink from your canteen.

Standards: Protect yourself from CB agent and/or unknown toxic industrial chemical contamination by donning, clearing, and checking your assigned protective mask within 9 seconds without becoming contaminated. Drink water through your protective mask from your canteen without becoming a casualty.

Special Condition: Do not wear contact lenses when performing this task. Do not use masks with damaged filters because certain models contain hazardous materials. Do not change filter elements in a contaminated environment.

Special Standards: Note: The mask gives you immediate protection against traditional warfare agents. The mask may not be adequate to protect you from certain toxic industrial chemicals, but it provides the best available protection to enable you to evacuate the hazard area. You may be required to evacuate to a minimum safe distance of at least 300 meters upwind from the contamination (if possible) or as directed by the commander.

Special Equipment:

Cue: If you hear or see a chemical or biological attack.

*Note:*Soldier must complete steps 1 through 4 within 9 seconds.

Performance Steps

WARNING
Before donning and adjusting the mask, female warfighters will remove earrings, hair fasteners (hair clips, hair pins, combs, and rubber bands), hair knots, buns, or braids that will interfere with the mask seal and will let hair hang freely. When wearing the Ground Crew Ensemble, hair will be neatly tucked inside jacket. Facial hair could result in an improper mask fit resulting in illness or death.   Do not wear contact lenses (soft or hard) while wearing the masks. Inadequate oxygen supply to the corneal surface, and exposure to dust, dirt, and smoke or gas may cause serious vision loss or eye damage. Personnel requiring vision correction will use the optical inserts that have been provided to them with their protective masks.

1. Don the mask.
   a. Stop breathing, and close eyes.
   b. Remove helmet, put helmet between legs above knees or hold rifle between legs and place helmet on the muzzle. If helmet falls continue to mask.

c. Take off glasses, if applicable.

d. Open the mask carrier with left hand.

e. Grasp the mask assembly with right hand, and remove it from the carrier.

f. Place chin in the chin pocket, and press the facepiece tight against face.

Note: The temple and forehead straps have already been adjusted during fitting.

g. Grasp the tab and pull the head harness over the head. Ensure that the ears are between the temple straps and the cheek straps. Ensure that the head harness is pulled far enough over the head that the forehead straps are tight.

h. Use one hand to tighten the cheek straps, one at a time, while holding the head pad centered on the back of head with the other hand. Ensure that the straps lay flat against the head.

2. Clear the mask.

a. Seal the outlet disk valve by placing one hand over the outlet valve cover assembly.

b. Blow out hard to ensure that any contaminated air is forced out around the edges of the facepiece.

3. Check the mask.

a. Cover the inlet port of the filter canister or the inlet port of the armor quick disconnect with the palm of the hand, and inhale.

b. Ensure that the facepiece collapses against face and remains so while holding your breath, which indicates that the mask is airtight.

c. Remove any hair, clothing, or other matter between the face and the mask if the facepiece does not collapse to face.

d. Notify the chemical, biological, radiological, and nuclear (CBRN) noncommissioned officer (NCO) if the mask still does not collapse.

4. Resume breathing.

Note: There is no time standards for donning the hood.

5. Completes steps 1 through 4 within 9 seconds.

CAUTION

BE CAREFUL WHEN PULLING ON THE HOOD BECAUSE IT COULD SNAG AND TEAR ON THE BUCKLES OF THE HEAD HARNESS.

---

6. Secure the mask hood.

WARNING

Be careful not to break facepiece seal when pulling protective hood over your head.

---

a. For the M50/M51-series protective mask.

(1) Place hands up under the protective hood, stretch elasticized portion and raise protective hood up
and over filters.

(2) Carefully pull excess protective hood material over head, neck and shoulders.

(3) Grasp underarm straps.

(4) Bring the male end of each underarm strap and fasten to female end.

(5) Tighten underarm straps.

    b. For the M48 series protective mask.

    (1) Carefully pull the back of the hood assembly over the head so hood covers the head, neck and shoulders.

    (2) Tuck inner skirt inside the collar of the CBRN protective suit. This can be done using the buddy system.

    c. For the M42- or M43-series protective mask, pull the hood over the head and zip the front closed to cover the bare skin.

    d. For the M45-series protective mask, pull the M7 hood over the helmet and head so that the hood covers shoulders.

    e. For the M40-series protective mask, don the hood so that it lies smoothly on the head.

    (1) For masks equipped with the regular hood—

    (a) Grasp the back edge of the hood skirt.

    (b) Pull the hood completely over the head so that it covers the back of the head, neck, and shoulders.

    (c) Zip the front of the hood closed by pulling the zipper slider downward.

    (d) Tighten the draw cord.

    (e) Secure the underarm straps by fastening and adjusting them.

    (f) Close your mask carrier.

    (g) Continue your mission.

    (2) For masks equipped with the quick-doff hood—

    (a) Place hands inside the hood and expand the elastic gathering around the neck of the hood.

    (b) Stretch and carefully pull the hood over head so that the hood covers your head, neck and shoulders.

    (c) Fasten the underarm straps.

    (d) Put on the helmet.

Note: For combat vehicle crewman (CVC) helmet, perform the following steps: 1) disconnect the boom microphone from the helmet, 2) connect the mask microphone to the receptacle in the helmet, 3) grasp the helmet next to the ear cups with the hand, and spread the helmet as far as possible, 4) place the helmet over head, tilting the helmet forward slightly so that the first contact when putting it on is with the forehead surface of the mask and 5) rotate the helmet back and down over the head until it is seated in position.

    (e) Close mask carrier.

    (f) Continue the mission.

Note: If the Soldier is using the mask in conjunction with the joint-service, lightweight integrated suit technology (JSLIST), he/she skips this step (the mask lacks a hood because it is built in on the JSLIST).

```
WARNING

USE M8 DETECTOR PAPER TO CHECK FOR CONTAMINATION
BEFORE USING THE DRINKING SYSTEM. IF CONTAMINATION IS
DETECTED, DECONTAMINATE USING THE M295
DECONTAMINATION KIT. DO NOT CONNECT THE QUICK-
DISCONNECT COUPLING TO YOUR CANTEEN UNTIL ALL
SURFACES ARE FREE OF CONTAMINATION. CHEMICAL
CONTAMINATION COULD ENTER YOUR MOUTH, AND YOU
COULD BECOME A CASUALTY.


DO NOT BREAK THE MASK SEAL WHILE DRINKING FROM THE
CANTEEN.
```

7. Drink water while wearing the mask.

a. Press in on the top of the outlet valve cover until the internal drink tube can be grasped between your teeth.

b. Steady the mask assembly with one hand and pull the quick-disconnect coupling out of the outlet valve cover. For the M50/M51 protective mask pull drink coupler out of coupler receptacle, below the front module.

c. Flip open the cover on the M1 canteen cap or open retaining strap on water canteen cap for the M50/M51 protective mask.

```
WARNING

IF RESISTANCE IS NOT FELT, YOUR DRINKING SYSTEM IS
LEAKING. DO NOT DRINK. REPLACE YOUR CANTEEN. IF
RESISTANCE IS STILL NOT FELT, NOTIFY YOUR CBRN NCO.
```

d. Push the quick-disconnect coupling into the canteen cap so that the pin enters the quick-disconnect coupling. For the M50/M51 protective mask push drink coupler into canteen cap so that seal snaps into the groove in the cap.

e. Turn drink tube lever on front module assembly upward, until it stops and is fully opened, to positioninternal drink tube in front of mouth, and grasp internal drink tube between your lips (for the M50/M51 and M48 protective mask only.)

```
WARNING

DO NOT TILT YOUR HEAD BACK WHILE DRINKING.
```

f. Blow to create positive pressure. You should feel some resistance.

g. Raise the canteen upside down and drink (if the system does not leak.)

h. Stop drinking after several swallows and lower the canteen. Blow into the internal drink tube to prevent the canteen from collapsing. Repeat the drinking procedure as required.

8. Doff the mask for storage.

a. M50/M51 protective mask

(1) Remove headgear.

(2) Loosen cheek straps completely by placing your thumbs behind the buckles and pulling forward so straps become loose. Grasp the front of the mask and lift it off your head.

(3) Replace headgear.

(4) Stow the audio frequency amplifier in the retaining loop located in the bottom of the mask carrier main stowage area before stowing the mask.

(5) Grasp the cheek straps and carefully pull the head harness over the front of the mask.

(6) Grasp the mask carrier flap tab and pull to open mask carrier flap.

(7) Grasp the mask by the front module assembly, place in mask carrier eyelens first, covered by head harness skullcap and face it away from the body.

b. M40A1 protective mask

(1) Remove helmet.

(2) Loosen cheek straps.

(3) Place one hand on the front voicemitter to hold mask assembly on face, with other hand grasp head harness tab, pull the head harness over the front of the mask assembly and remove mask assembly.

(4) Replace helmet.

(5) Pull head harness over front of mask assembly.

(6) Smooth the second skin/universal second skin over the front of the mask assembly.

(7) Pull the forehead straps tight over the second skin/universal second skin, by pulling the head harness down as far as possible, by pulling on the harness tab.

(8) Hold the facepiece assembly up and put it in the mask carrier with the lenses facing away from your body.

CAUTION

It is important to completely close the hook and pile fastener on the mask carrier cover. Failure to do this will result in collection of debris and damage to the mask assembly.

---

(9) Close the mask carrier. Seal the entire hook and pile fastener surface.

(10) Stowing M40A1/M42-Series Mask With Quick Doff Hood (QDH).

(a) Hold front of mask assembly in a horizontal position and smooth the QDH over it.

(b) Store the ends of the underarm straps in a "V".

(c) Fold the two edges of the QDH over the underarm straps to create a "V".

(d) Fold the "V" up to cover the eyelenses. Do not let the QDH cover the chin opening.

(1) If stowing M42-series, wrap the two edges of the hood over the underarm straps and around the hose.

(2) Grasp the hose through the hood and aligned hose to point down toward the chin.

     (e) Hold the facepiece assembly up and put it in the mask carrier with the lenses facing away from your body.

     (f) Close the mask carrier. Seal the entire hook and pile fastener surface.

    c. M43 protective mask

     (1) Remove helmet.

     (2) Detach canister disconnect along with canisters from top of blower. Let hose and canisters hang from facepiece.

     (3) Turn blower off.

     (4) Using both hands, gently lift hood up over head; let hood hang from front of facepiece.

     (5) Loosen three head harness straps by rolling buckles forward.

     (6) Grip head harness and pull facepiece up and off head.

     (7) Check that the facepiece is dry and free of oil or solvents before stowing. If facepiece is not dry and free of oil or solvents, clean as directed.

     (8) With hood hanging in front of facepiece, hold front of facepiece in a horizontal position (face down) and smooth the hood beneath it.

     (9) Fold hood and microphone cable up around the back of facepiece.

     (10) Open top flap on carrier.

     (11) To stow the items in the proper locations proceed as follows:

     (a) Place hose, canister disconnect and canisters into carrier section on the nametag side of the carrier.

     (b) Place facepiece on top of canisters.

     (c) Unsnap blower from harness.

     (d) Insert blower with controls up, into carrier section on the back side of the carrier.

     (e) Unclip shoulder, waist, leg straps and remove blower harness.

     (f) Stow blower harness in carrier section with blower.

     (g) Make sure technical manual is stowed in carrier section with blower.

     (h) Close and seal top flap on carrier.

     (i) Store carrier with contents in dark, dry location. Hang carrier by one of the straps for storage.

    d. M45 protective mask.

     (1) Remove helmet.

     (2) Disconnect microphone (if issued) from helmet receptacle, then remove helmet.

     (3) Loosen cheek straps ONLY.

     (4) Place thumbs under both cheek tabs. Lift bottom of mask out and up over your head.

     (5) Pull head harness over front of mask.

     (6) With canister toward the bottom of the carrier, place mask in carrier with eyelenses facing up and away from body.

     (7) Close carrier.

e. M48 protective mask

(1) Remove helmet.

(2) Using both hands, gently lift hood of facepiece assembly up over head; let hood hang from front of facepiece assembly.

(3) Loosen the three suspension harness straps by rolling buckles forward.

(4) Grip suspension harness and pull facepiece assembly up and off head.

(5) Turn blower assembly off.

WARNING

Any occurrences of redness, puffiness, or itchiness that persist for an extended period of time after removing the facepiece assembly should be referred to the flight surgeon for evaluation.

---

(6) Perform all After PMCS

(7) With hood assembly hanging in front of facepiece assembly, install facepiece assembly in facepiece carrier with open side of facepiece assembly toward the leg (back of facepiece carrier.)

(8) Tuck remainder of hood into facepiece carrier. Leave hose exposed from facepiece carrier.

(9) Close flap on facepiece carrier.

9. Perform all steps in sequence without becoming a casualty.

Evaluation Preparation:

Setup: Evaluate this task during a field exercise or a tactical training session. Use a mask earlier fitted to the Soldier's face. The Soldier will bring his/her flight or CVC helmet. The Soldier should be in mission-oriented protective posture 4 (MOPP4). Do not use a new decontamination kit for every Soldier; use the kit as long as possible. Ensure that the Soldier has M8 detector paper in the mask carrier before testing.

Brief Soldier: Tell the Soldier to stand, while wearing his/her mask carrier containing his/her assigned protective mask with the hood attached. Provide the Soldier with one of the scenarios described in the conditions statement (cue to begin masking). Tell the Soldier to keep the mask on until you issue the all clear order. Tell the Soldier to drink water while wearing his/her assigned mask.

Performance Measures	GO	NO GO
1 Donned the mask.	____	____
2 Cleared the mask.	____	____

Performance Measures	GO	NO GO
3 Checked the mask.	_____	_____
4 Resumed breathing.	_____	_____
5 Completed steps 1 through 4 in sequence within 9 seconds.	_____	_____
6 Secured the mask hood.	_____	_____
7 Consumed water while wearing the mask.	_____	_____
8 Doffed the mask for storage.	_____	_____
9 Performed all steps in sequence without becoming casualty.	_____	_____

Evaluation Guidance: Score the Soldier GO if all performance measures are passed (P). Score the Soldier NO-GO if any performance measure is failed (F). If the Soldier fails any performance measure, show him how to do it correctly.

References:
Required: FM 3-11.3, FM 3-11.4, TM 3-4240-300-10-2, TM 3-4240-312-12&P, TM 3-4240-346-10, TM 3-4240-542-13&P
Related:

031-COM-1019

React to Chemical or Biological (CB) Hazard/Attack

Conditions: You are in a tactical environment where the threat of an attack or exposure to chemical or biological agents (toxic industrial or conventional warfare) is high. You are given mission oriented protective posture (MOPP) gear, protective mask, an individual decontamination kit, reactive skin decontamination lotion (RSDL) and M8 and M9 detector paper. You may also be given eye protection, an Army combat helmet (ACH), an improved outer tactical vest (IOTV), and deltoid auxiliary protectors (DAPs). You are now in MOPP 0 and one of the following automatic-masking situations occurs: 1) a chemical alarm sounds, 2) a positive reading is obtained on detector paper, 3) individuals exhibit symptoms of chemical biological (CB) agents or toxic industrial chemical (TIC) poisoning, such as difficulty breathing, coughing, wheezing, vomiting, or eye irritation, 4) you observe a spill or cloud of unknown material(s), 5) you react to an improvised explosive device (IED) explosion where you suspect the release of a CB or toxic chemical, 6) you observe a contamination marker, 7) your supervisor orders you to mask, 8) you observe personnel wearing protective masks, and 9) you observe other signs of a possible CB agent or toxic industrial chemical attack/spill.

Standards: React to a CB agent or toxic industrial chemical hazard/attack without becoming a casualty by: donning your protective mask within 9 seconds, notifying your supervisor of identified or possible contamination, starting the steps to decontaminate yourself within 1 minute of finding contamination, (after decontaminating yourself all over) assuming MOPP 4 unless directed to a lower MOPP level, decontaminating your individual equipment using the decontaminating kit as necessary, and continuing the mission.

Special Condition: None

Special Standards: None

Safety Risk: Low

Cue: None

Note: None

Performance Steps

1. Identify the CB hazard automatic-masking criteria.
   a. Don your protective mask automatically when any of the following situations occur:
      (1) A chemical alarm sounds.
      (2) A positive reading is obtained on detector paper.
      (3) Individuals exhibit symptoms of CB agent poisoning, such as difficulty breathing, coughing, wheezing, vomiting, or eye irritation.
      (4) You observe a spill or cloud of unknown material(s).
      (5) You react to an IED explosion where you suspect the release of a CB agent.
      (6) You observe a contamination marker.
      (7) Your supervisor orders you to mask.
      (8) You observe personnel wearing protective masks.

(9) You observe other signs of a possible CB agent hazard/attack.

    b. Respond to the commander's policy of automatic masking.

Note: Commanders at all levels may establish a modified policy by designating additional criteria for automatic masking.

    2. Protect yourself from CB contamination by using your assigned protective mask without fastening the hood within 9 seconds.

Note: The mask provides protection against conventional warfare agents. The mask provides little if any protection from toxic industrial materials (TIMs), but it provides the best available protection to enable you to evacuate the hazard area. You may be required to evacuate to a minimum safe distance of at least 300 meters upwind from the contamination (if possible) or as directed by the commander.

    a. Stop breathing and close your eyes.

    b. Don the protective mask with hood.

    c. Clear the mask.

    d. Check the mask.

    e. Do not fasten the hood.

    f. Go immediatley to the next step.

    3. Give the alarm.

    a. Shout, "Gas, Gas, Gas."

    b. Give the appropriate hand-and-arm signal.

    c. Hit two metal objects together.

    4. Take cover and/or assemble as directed, moving at least 300 meters upwind from the suspected contamination area to reduce exposure.

    5. Decontaminate exposed skin within 1 minute of becoming contaminated using the individual decontaminating kit as necessary.

    6. Cover all exposed skin and assume MOPP 4 as directed.

Note: This step is graded only if MOPP is available.

Note: If you are wearing an ACH, IOTV, or DAPs, proceed to performance step 6e through 6k.

    a. Don the overgarment trousers.

    b. Don the overgarment coat.

    c. Don the overboots.

Note: Combat boots provide limited protection. Cover them as soon as possible because they absorb chemicals. (It takes a long time to put on the overboots; in an emergency, put them on last.)

    d. Don the protective gloves.

    e. Remove the ACH and protective eyewear.

    f. Loosen the DAPs.

WARNING

WHEN DOFFFING THE IOTV FROM THE SHOULDER. TAKE CARE NOT TO SNAG THE FILTER CANISTER AND BREAK THE SEAL OF YOUR PROTECTIVE MASK.

---

    g. Doff the IOTV by lifting the front flap and detach side plate carriers by separating hook and loop fastener tape. Lift front carrier and detach internal elastic bands at hook and loop interface. Open the medical access hook and pile closure, loosen the left shoulder adjustment strap and slide vest off the right shoulder.

     h. Perform performance steps 6a through 6d, and then proceed to performance step 6i.

     i. Don the IOTV over the right shoulder by tightening the left shoulder adjustment strap and fastening the medical access hook and pile closure. Attach internal elastic bands at hook and loop interface and close the front carrier. Attach side plate carriers and close the front flap.

     j. Secure the DAP.

     k. Don the ACH.

   7. Decontaminate your individual equipment using your individual equipment decontamination kit, as necessary.

   8. Notify your supervisor of any suspected CB hazard/attack.

   9. Continue the mission and perform any additional requirements as outlined in your unit's standing operating procedure (SOP).

Note:

1. Use all means of CB detection to check your surrounding area for the presence of contamination.

2. Contact your higher headquarters if you find contamination or if you determine that the attack was non-CB related.

3. Await further guidance. The higher headquarters contacts all adjacent/attached units to check the status of CB contamination in their areas. All units will report the absence or presence of contamination to the chain of command.

4. Annotate the above actions on your duty log Department of the Army (DA) Form 1594, Daily Staff Journal or Duty Officer's Log, and as a siginificant activity (SIGAct) on units Combat Management System (CMS).

Evaluation Preparation:

Setup: A good time to evaluate this task is during a field exercise when a variety of CB hazards can be simulated. Select a site with adequate cover, and ensure that Soldiers have their assigned protective mask.

Brief Soldier: Tell the Soldier that there will be an encounter with simulated CB agents and/or a CB alarm will be given

Performance Measures	GO	NO GO
1 Identified automatic-masking criteria.	_____	_____

Performance Measures	GO	NO GO
2 Donned protective mask without fastening the hood within 9 seconds.	_____	_____
3 Gave the alarm.	_____	_____
4 Took cover and/or assembled as directed. Moved at least 300 meters upwind from the suspected contamination area to reduce exposure.	_____	_____
5 Decontaminated exposed skin as necessary within 1 minute of finding the contamination.	_____	_____
6 Covered all exposed skin and assumed MOPP 4 as directed. (Graded only if MOPP is available. If Soldier is wearing an ACH, IOTV, or DAPs, proceed to performance step 6e and 6K.)	_____	_____
7 Decontaminated individual equipment as necessary.	_____	_____
8 Notified the supervisor of any CB hazard/attacks.	_____	_____
9 Continued the mission and performed requirements outlined in unit's SOP.	_____	_____

Evaluation Guidance: Score the Soldier GO if all performance measures are passed (P). Score the Soldier NO-GO if any performance measure is failed (F). If the Soldier fails any performance measure, show him how to do it correctly.

References:
Required: DA FORM 1594, FM 3-11.4, TM 3-4230-229-10, TM 3-4230-235-10
Related:

References:

031-COM-1040

Protect Yourself from CBRN Injury/Contamination with the JSLIST Chemical-Protective Ensemble

Conditions: You are given the Joint-Service, Lightweight, Integrated Suit Technology (JSLIST) chemical-protective ensemble consisting of JSLIST overgarments (coat and trousers), JSLIST compatible protective mask, JSLIST compatible footwear covers, and JSLIST compatible protective gloves, Field Manual (FM) 3-11.4, Technical Manual (TM) 10-8415-220-10, and Skin Exposure Reduction Paste Against Chemical Warfare Agents (SERPACWA). You are directed to assume mission-oriented protective posture (MOPP) level 4, or you are in a situation where you must automatically react to a chemical, biological, radiological, and nuclear (CBRN) hazard.

Standards: Protect yourself from CBRN injury/contamination with the JSLIST chemical protective ensemble by: 1) performing Before preventive maintenance checks and services (PMCS) according to TM 10-8415-220-10, 2) applying SERPACWA if command directed, 3) assuming MOPP Levels 1 through 4 in order within 8 minutes, 4) doffing the JSLIST ensemble, 5) performing After PMCS inspections, and 6) repackaging the JSLIST ensemble.

Special Condition: None

Special Standards: None

Special Equipment:

Cue: You are ordered to do so, learn a chemical attack is about to happen, must enter an area where chemical agents have been used, recognize a chemical hazard, or attacked with chemical agents without warning.

Note: None

Performance Steps

1. Perform Before PMCS on the JSLIST ensemble in accordance with (IAW) TM 10-8415-220-10.
   a. Using a new garment, first use.
     (1) Remove the coat or trousers from the factory vacuum-sealed bags and store in trouser pocket.

Note: Coat and trousers packaging includes resealable bags. Store bags in trousers pockets and retain for reuse in repackaging the JSLIST ensemble.

 (2) Perform Before PMCS according to table 2-1 located in TM 10-8415-220-10.

 (3) Mark the label with the date that the garment was removed from the package in permanent ink.

 b. Using a used garment.

  (1) Remove the coat or trousers from the clear plastic, resealable bag.

  (2) Check the wear date marked on the labels. If more than 120 days have elapsed since this date, replace the coat or trousers with new coat or trousers.

Note: To conserve chemical protective overgarment assets, any used JSLIST overgarment coat or trouser that has exceeded the 120-day period may be used as a training only item (the words TRAINING ONLY must be stenciled 2.5 inches high or larger on the outside of a sleeve or leg of the item, in a contrasting colored permanent ink).

WARNING

SERPACWA is for military and external use only. Do not apply to the eyes or to mucous membranes. This product, its packaging, and clothing or other materials exposed to SERPACWA should not be destroyed by burning due to the release of toxic fumes. Smoking should be avoided, be sure to avoid getting SERPACWA on smoking products. Be sure to clean hands before handling smoking products.

---

 2. Apply SERPACWA (if command-directed).

Note: SERPACWA is intended for use prior to exposure to chemical warfare agent (CWA) and only in conjunction with the JSLIST chemical protective ensemble.

 a. Before you assume MOPP Level 1, use a dry towel to wipe off sweat, insect repellent, camouflage paint, sand, or dirt from your skin at the areas shown on the packet label.

 b. Tear open the packet and squeeze about one-third of the pouch into the palm of your hand and rub it evenly around the wrists (site 1), neck (site 2), and boot tops of lower legs (site 3) until it forms a difficult to notice white film.

 c. Remove the remaining two-thirds of the SERPACWA from the pouch and rub it evenly onto your armpits (site 4), groin area (site 5), and waistline (site 6).

 d. After SERPACWA is applied, if exposure to CWA is either confirmed or suspected, follow the appropriate protocol for decontamination.

 e. For removal of SERPACWA in the absence of exposure to CWA, scrub the sites with a dry towel, or if possible, with a cloth using both soap and water.

 3. Don the JSLISTS chemical protective ensemble, in MOPP Level 1 through 4 sequence within eight minutes.

 a. Assume MOPP Level 1.

  (1) Don the JSLIST overgarment trousers.

   (a) Extend your toes downward, put one leg into the trousers, and pull them up. Repeat the procedure for your other leg.

(b) Close the slide fastener, and fasten the two fly opening snaps.

(c) Pull the suspenders over your shoulders, and fasten the snap couplers. Adjust the suspenders to ensure that the trousers fit comfortably.
Note: The trouser length can be adjusted by raising or lowering the suspenders.

(d) Adjust the waistband hook-and-pile fasteners for a snug fit.

(2) Don the JSLIST overgarment coat.

(a) Don the coat, and close the slide fastener up as far as your chest.

(b) Secure the front closure hook-and-pile fasteners up as far as your chest.

(c) Pull the bottom of the coat down over the trousers. Pull the loop out and away from the overgarment coat, and bring it forward between the legs. Pull on the loop until the bottom of the coat fits snugly over the trousers.

b. Assume MOPP Level 2. Don the overboots.

(1) Don the overboots (multipurpose overboots/black vinyl overboots/green vinyl overboots (MULO/BVO/GVO) over the combat boots. Adjust and secure the strap-and-buckle fasteners.

(2) Pull the trouser legs over the overboots (MULO/BVO/GVO). Secure the hook-and-pile fasteners on each ankle to fit snugly around the boot.

c. Assume MOPP Level 3 by donning chemical-protective mask IAW task 031-503-1035.

d. Assume MOPP Level 4. Don the gloves.

(1) Push the sleeve cuffs up your arm.

(2) Put on the gloves and glove liners (inserts).

(3) Pull the sleeve cuffs over the top of the gloves, and secure the hook-and-pile fastener tape snugly on each wrist.

4. Doff the JSLIST chemical protective ensemble.

a. Doff the gloves.

(1) Unfasten the hook-and-pile fastener tape on each wrist, and remove the gloves (and liners if butyl rubber gloves are worn).

(2) Put the gloves in the trouser pockets.

b. Untie the bow in the coat retention cord, unfasten the webbing strip snap, and release the coat retention cord loop.

c. Doff the helmet and cover, if worn.

d. Doff the hood from the JSLIST coat.

(1) Unfasten the barrel locks.

(2) Loosen the hood.

(3) Unfasten the hook-and-pile fastener tape at your neck.

(4) Pull the hood off your head.

e. Doff the protective mask, stow it in the carrier, remove the carrier, and place on an uncontaminated surface.

f. Doff the overboots.

(1) Unfasten the ankle hook-and-pile fastener tapes.

(2) Unfasten the two strap-and-buckle fasteners on the MULOs.

(3) Remove the MULOs.

g. Doff the JSLIST coat.

(1) Unfasten the front closure flap hook-and-pile fastener tape and the front slide fastener.

(2) Remove the coat.

h. Doff the JSLIST trousers.

(1) Unfasten the suspender and waist hook-and-pile fastener tapes.

(2) Unfasten the front closure snaps, and open the slide fastener.

(3) Remove the JSLIST trousers.

5. Perform After PMCS inspections according to table 2-1 and paragraph 3-2 in TM 10-8415-220-10.

6. Repackage the JSLIST ensemble.

a. Remove the clear plastic, resealable bags from the trousers pocket.

b. Fold and repack the coat and trousers in individual clear plastic, resealable bags.

Evaluation Preparation:

Setup: Provide the Soldier with the items listed in the task conditions statement. Evaluate this task during a field exercise or during a normal training session. Ensure that adequate amounts of serviceable JSLIST ensembles are present or that the evaluated Soldier(s) bring their JSLIST ensemble to the evaluation site. The evaluator must be prepared to direct higher MOPP Levels at once as a Soldier reaches a preceding Level.

Brief Soldier: Read the action, conditions, and standards to the Soldier. Tell the Soldier to complete before PMCS and to inform you of any faults found with the suit (for example, the draw cord is unserviceable, and so forth). After the before PMCS has been completed, inform the Soldier that MOPP Levels 1 through 4 must be achieved in sequence and that the time limit for achieving MOPP Level 4 is 8 minutes. Ensure that the Soldier informs you as each MOPP Level is obtained (for example, "I am now at MOPP Level 1."). For performance measures or parts of performance measures requiring the Soldier to state an answer, lead the Soldier with an appropriate question (for example, "When would you put SERPACWA on?", or "What do you use the clear plastic, resealable bags for?").

Performance Measures	GO	NO GO
1 Performed Before PMCS inspection in accordance with (IAW) TM 10-8415-220-10.	___	___
2 SERPACWA was applied IAW the package instructions, if use was command directed.	___	___

Performance Measures	GO	NO GO
3 Donned the JSLIST chemical protective ensemble, in MOPP Level 1 through 4 sequence, within 8 minutes.	_____	_____
4 Doffed the JSLIST chemical protective ensemble.	_____	_____
5 Performed After PMCS IAW TM 10-8415-220-10. (Labeled the coat and trouser in permanent ink with the days of wear, and the number of times they had been laundered IAW unit SOP.)	_____	_____
6 Repackaged JSLIST ensemble. Removed the clear-plastic resealable bags from the JSLIST trouser pocket, and placed the used JSLIST coat and trousers in them and resealed.	_____	_____

Evaluation Guidance: Score the Soldier GO if all performance measures are passed (P). Score the Soldier NO-GO if any performance measure is failed (F). If the Soldier fails any performance measure, show him how to do it correctly.

References:
Required: FM 3-11.4, TM 10-8415-220-10
Related:

031-COM-1013

Decontaminate Yourself and Individual Equipment Using Chemical Decontaminating Kits

Conditions: You are at mission-oriented protective posture (MOPP) Level 2. You are given Technical Manual (TM) 3-4230-229-10, TM 3-4230-235-10, a chemical protective mask, chemical protective gloves, chemical protective overboots, a full canteen of water, a poncho, load-bearing equipment (LBE) or load-bearing vest, Interceptor Body Armor (IBA), the Improved Outer Tactical Vest (IOTV), and M291 decontaminating kit(s) or the Reactive Skin Decontamination Lotion (RSDL). Your skin and eyes have been exposed to chemical agents, or you have passed through a chemically contaminated area and suspect that you have contamination on your skin.

Standards: Decontaminate yourself and your individual equipment using the chemical decontaminating kits. Start the steps to decontaminate your skin and eyes within 1 minute after contamination. Decontaminate your exposed skin and eyes, as necessary, before chemical-agent symptoms occur. Decontaminate all individual equipment after decontaminating your skin and eyes.

Special Condition: None

Special Standards: None

Special Equipment:

Cue: None

*Note:*None

Performance Steps

 1. Assume MOPP Level 3 without securing the hook-and-pile fastener tape or drawcord on the integrated hood.
Note: For training purposes, use the Training RSDL. If the Reactive Skin Decontaminating Lotion (RSDL) is not available, use the M291 (skip to performance step 3).
 2. Decontaminate your skin using the RSDL within 1 minute of contamination.

DANGER
Do not mix RSDL with solid, undiluted high-test hypochlorite (HTH) or super tropical bleach (STB). Heat and/or fire may result.

WARNING
Under no circumstances should the training RSDL be used in place of the RSDL during actual combat operations. The training lotion does not contain active ingredients.

   a. Decontaminate your hands, face, and the inside of your mask.
      (1) Remove one RSDL packet from your carrying pouch.
      (2) Tear it open quickly at any notch.
      (3) Remove the applicator pad from the packet, and save the packet as the remaining lotion can be added to the applicator pad, if required.
      (4) Thoroughly scrub the exposed skin of your hand, palm, and fingers with the applicator pad.
Note: The applicator pad can be used from either side and may gripped in any manner allowing the applicator pad to be applied to the skin.
      (5) Switch the applicator pad to the other hand, and repeat the procedure.

DANGER

Death or injury may result if you breathe toxic agents while doing the following steps. If you need to breathe before you finish, reseal your mask, clear it, check it, get your breath, and then resume the decontaminating procedure.

(6) Stop breathing, close eyes, grasp mask beneath chin and pull mask away from chin enough to allow one hand between the mask and your face. Hold the mask in this position during steps 2a(7) through 2a(13).

Note:

1. Do not discard the applicator pad at this time.
2. If you were masked with your hood secured when you became contaminated, stop. Put on your protective gloves, and proceed to step 2b.
3. If you were not masked with the hood secured when you became contaminated, continue decontaminating the exposed skin.

(7) Thoroughly scrub the exposed skin of your face with lotion from the applicator pad.

(8) Thoroughly scrub across your forehead.

(9) Beginning at one side, scrub up and down across your cheeks, nose, chin, and closed mouth. Avoid ingesting.

(10) Scrub under the chin from the ear along the jawbone to the other ear to coat your skin with lotion

CAUTION

Do not apply lotion to the lens of the protective mask. The RSDL may cause loss of transparency.

(11) Turn your hand over and scrub the inside surfaces of the mask that may touch your skin. Be sure to include the drinking tube.

(12) Keep the applicator.

(13) Seal your mask immediately, clear it, and check it.

(14) Use the applicator and any remaining lotion in the packet. Without breaking the mask seal, scrub the applicator pad across the forehead, exposed scalp, the skin of the neck, ears, and throat.

(15) Secure the hood.

(16) Thoroughly scrub your hands with lotion again as in steps 2a(4) through 2a(5).

(17) Assume MOPP Level 4 by putting on protective gloves.

WARNING

Do not discard the RSDL packaging or applicator pads into containers that contain HTH or STB. Heat and/or fire may result.

b. Use any remaining lotion to spot decontaminate weapons, personal equipment, and canteen cap that may have become contaminated.

c. Allow RSDL to remain on skin for at least 2 minutes to destroy the chemical agent.

d. Discard the used packet(s) and applicator pad(s) by leaving them in place.

Note: Do not put used packets in your pockets. Discard the carrying pouch after using the packets.

 e. Remove the decontaminating lotion with soap and water when operational conditions permit, such as an "All Clear" directive or after detailed troop decontamination.

Note: Upon completion of training and evaluation, ensure that Soldiers have adequate mask cleaning supplies and water to clean training RSDL off of their protective mask.

 3. Decontaminate your skin using the M291 decontaminating kit within 1 minute of contamination.

DANGER

Death or injury may result if you breathe toxic agents while decontaminating your face. If you need to breathe before you finish, reseal your mask, clear it, check it, get your breath, and then resume the decontaminating procedure.

CAUTION

The M291 decontaminating kit is for external use only. Keep decontaminating powder out of your eyes and out of any cuts or wounds. The decontaminating powder may irritate your skin or eyes.

If your face has been contaminated, use water to wash the toxic agent out of your eyes, cuts, or wounds.

After decontaminating with water, cover exposed cuts or wounds with appropriate first aid wrap or bandages before handling the decontaminating kit.

Do not handle or hold leaking packets above your head. Do not touch or rub your eyes, lips, or the inside of your mouth with anything that has been in contact with the decontaminating powder.

Do not attempt to decontaminate a loaded weapon. Always unload and clear the weapon and place the weapon on safe before starting decontaminating procedures. Immediate decontaminating techniques remove only the liquid hazard. Certain items may still present a vapor hazard. See your supervisor for unmasking procedures.

(1) Remove one skin decontamination packet from your carrying pouch.

(2) Tear it open quickly at the notch.

(3) Remove the applicator pad from the packet, and discard the empty packet.

(4) Unfold the applicator pad, and slip your finger(s) into the handle.

(5) Scrub the back of your hand, palm, and fingers until they are completely covered with black powder from the applicator pad.

(6) Switch the applicator pad to the other hand, and repeat the procedure.

Note:

1. Do not discard the applicator pad at this time.

2. If you were masked with your hood zipped and the drawstring pulled tight when you were contaminated, stop. Discard the applicator pad, put on your protective gloves, and go to step 3b. However, if you were masked, but the zipper and drawstring were not secure, go to step 3a(16). The stars in the illustration on page 2-5 of TM 3-4230-229-10 show areas of the face that

should be scrubbed with an extra stroke because they are hard to decontaminate.

3. The procedure is the same regardless of the type of protective mask. If you are using the Joint Service Lightweight Integrated Suit Technology (JSLIST) with a hood attached to the protective jacket, ignore the instructions for the hood.

> **DANGER**
>
> Death or injury may result if you breathe toxic agents while decontaminating your face. If you need to breathe before you finish, reseal your mask, clear it, check it, get your breath, and then resume the decontaminating procedure.

(7) Scrub exposed skin of your face thoroughly until you are completely covered with black powder from the applicator pad.

(8) Hold your breath, close your eyes, grasp the mask beneath your chin, and pull the hood and mask away from your chin enough to allow one hand between the mask and your face.

(9) Scrub up and down across your face, beginning at the front of one ear, to your nose, and then to your other ear.

(a) Scrub across your face to the corner of your nose.

(b) Scrub an extra stroke at the corner of your nose.

(c) Scrub across your nose, to the tip of your nose, and then to the other corner of your nose.

(d) Scrub an extra stroke at the corner of your nose.

(e) Scrub across your face to your other ear.

(10) Scrub up and down across your face to your mouth and then to the other end of your jawbone.

(a) Scrub across your cheek to the corner of your mouth.

(b) Scrub an extra stroke at the corner of your mouth.

(c) Scrub across your closed mouth to the center of your upper lip.

(d) Scrub an extra stroke above your upper lip.

(e) Scrub across your closed mouth to the outer corner of your mouth

(f) Scrub an extra stroke at the corner of your mouth.

(g) Scrub across your cheek to the end of your jawbone.

(11) Scrub up and down across your face to your chin and then to the other end of your jawbone.

(a) Scrub across and under your jaw to your chin, cupping your chin.

(b) Scrub extra strokes at the center of your chin.

(c) Scrub across your upper jaw to the end of your jawbone.

(12) Turn your hand out, and quickly wipe the inside of your mask where it touches your face.

(13) Discard the applicator pad.

(14) Seal your mask immediately, clear it, and check it.

(15) Remove the second skin decontamination packet from the carrying pouch.

(16) Repeat steps 3a(2), (3), and (4) above.

(17) Scrub your neck and ears until they are thoroughly covered with black powder without breaking the seal between your face and your mask. Scrub your hands again until they are completely covered with black powder.

    b. Assume MOPP Level 4.

       (1) Discard the applicator pad.

       (2) Put on your protective gloves.

       (3) Fasten your hood.

    c. Remove the decontaminating powder with soap and water when operational conditions permit.

  4. Decontaminate your individual equipment using the M295 decontaminating kit.

    a. Use the first mitt to decontaminate your gloves, the exposed areas of your mask and hood, your weapon, and your helmet.

       (1) Remove one decontamination packet from your pouch.

       (2) Tear the packet open at any notch.

       (3) Remove the decontamination mitt.

       (4) Discard the empty packet.

       (5) Unfold the decontamination mitt.

       (6) Grasp the green (nonpad) side of the decontamination mitt with your nondominant hand. Pat the other gloved hand with the decontamination mitt to start the flow of decontamination powder onto your glove. Rub your glove with the decontamination mitt until it is completely covered with decontaminating powder.

       (7) Insert the decontaminated, gloved hand inside the decontamination mitt. Ensure that the pad side is in the palm of your hand and your thumb sticks through the appropriate thumbhole. Securely tighten the wristband on the gloved hand.

       (8) Decontaminate individual equipment by rubbing with the pad side of the decontamination mitt until the equipment is thoroughly covered with decontamination powder. Pay special attention to areas that are hard to reach (such as cracks, crevices, and absorbent materials).

         (a) Decontaminate your other glove.

         (b) Decontaminate exposed areas of your mask and hood.

         (c) Decontaminate your weapon.

         (d) Decontaminate your helmet by patting it with the decontamination mitt.

       (9) Discard the decontamination mitt.

    b. Use the second mitt to decontaminate your LBE, IBA or IOTV and accessories, mask carrier, overboots, and gloves again.

       (1) Get another packet, and repeat steps 4a(1) through 4a(7). Then, perform the following:

         (a) Decontaminate load-carrying equipment (LCE), IBA, IOTV and accessories (such as canteen, ammo pouch, and first aid pouch).

         (b) Decontaminate your mask-carrying case.

         (c) Decontaminate your protective boots.

         (d) Repeat the decontamination process on your protective gloves.

       (2) Discard the decontamination mitt.

(3) Get another packet and repeat steps 4a(1) through 4a(7) if liquid contamination is still suspected or detected. Rub or blot areas where contamination is still suspected or detected.

> **WARNING**
>
> The M295 kit only removes the liquid hazard. Decontaminated items may still present a vapor hazard. Do not unmask until it has been determined safe to do so.

    c. Remove the decontaminating powder when operational conditions permit.

    5. Notify your supervisor on the location of the used decontaminating materials, and await guidance on disposal procedures.

Evaluation Preparation:

Setup: Provide the Soldier with the items listed in the task conditions statement. A good time to evaluate this task is while in a field environment. Gather materials for the disposal of hazardous waste according to federal, state, and local rules and regulations.

Brief Soldier: Tell the Soldier what body parts and equipment are contaminated.

Performance Measures	GO	NO GO
1 Assumed MOPP Level 3 without securing the hook-and-pile fastener tape or drawcord.	____	____
2 Decontaminated skin using the RSDL within 1 minute of contamination.	____	____
3 Decontaminated skin using the M291 decontaminating kit within 1 minute of contamination.	____	____
4 Decontaminated individual equipment using the M295 decontaminating kit.	____	____

Performance Measures	GO	NO GO
5 Notified supervisor on the location of the used decontaminating materials, and awaited guidance on disposal procedures.	_____	_____

Evaluation Guidance: Score the Soldier GO if all performance measures are passed (P). Score the Soldier NO-GO if any performance measure is failed (F). If the Soldier fails any performance measure, show him how to do it correctly.

References:
Required: FM 3-11.4, FM 3-11.5, TM 10-8415-209-10, TM 10-8415-220-10, TM 3-4230-229-10, TM 3-4230-235-10
Related:

031-COM-1037

Detect Chemical Agents Using M8 or M9 Detector Paper

<table>
<tr><td align="center">WARNING</td></tr>
<tr><td>Always wear protective gloves when touching M9 detector paper. Do not get M9 detector paper in or near your mouth or on your skin. The M9 detector paper dye may cause cancer, but the risk is small because very little dye is used.</td></tr>
</table>

Conditions: You are in mission-oriented protective posture (MOPP) 2 in a tactical environment or an area where there is a chemical threat. You are given a protective mask, a booklet of M8 detector paper, a dispenser of M9 detector paper, M256A1 or M256A2 chemical-agent detector kit, assigned M291 skin decontaminating kit or reactive skin decontamination lotion (RSDL), M295 individual equipment decontamination kit, DA Form 1594 Daily Staff Journal or Duty Officer's Log, FM 3-11.4, FM 3-11.3, TM 3-6665-311-10, and a complete set of MOPP gear or a chemical-protective ensemble.

Standards: Detect chemical agents using M8 and M9 detector paper, ensuring that the M9 detector paper is attached to places likely to come into contact with liquid chemical agents. Detect and identify all liquid chemical agents in the area that are within the capabilities of the M8 or M9 detector paper without becoming a casualty.

Special Condition: 1) Do not wear contact lenses when performing this task, 2) do not use masks with damaged filters because certain models contain hazardous materials, and 3) do not change the filter element in a contaminated environment.

Special Standards: None

Special Equipment:

Cue: None

Note: None

Performance Steps

1. Detect chemical agents using M9 detector paper.
Note: M8 and M9 detector paper will not detect chemical-agent vapors.

    a. Attach the M9 detector paper to your MOPP gear and equipment while wearing chemical-protective gloves.

       (1) Place the M9 detector paper on the MOPP gear on opposite sides of your body.

          (a) If you are right-handed, place a strip of M9 detector paper around your right upper arm, left wrist, and right ankle.
Note: These are the places where a moving Soldier will most likely brush against a surface (such as undergrowth) that is contaminated with a liquid chemical agent.

          (b) If you are left-handed, place a strip of M9 detector paper around your left upper arm, right wrist, and left ankle.
Note: Do not attach M9 detector paper to hot, dirty, oily, or greasy surfaces because it may give a false positive reading.

       (2) Place M9 detector paper on equipment where it will come in contact with contaminated objects and is visible to the operator.

> **CAUTION**
>
> Firing weapons lubricated with lubricating oil, semi-fluid; lubricant, small arms; or lubricant, semifluid, automatic weapons (LSA) may cause false positive responses on the olive drab (OD) detector paper.

    b. Monitor the M9 detector paper constantly for any color change. If you observe a color change, immediately do the following:

       (1) Mask.

       (2) Give the alarm.

       (3) Decontaminate as necessary.

       (4) Assume MOPP 4.

2. Detect chemical agents using M8 detector paper if you see a liquid that might be a chemical agent or if you observe a color change on the M9 detector paper.

    a. Assume MOPP 4 immediately.

    b. Prepare the M8 detector paper. Tear out a sheet from the book (use one-half sheet if it is perforated).
Note: You may want to put the paper on the end of a stick or another object and then blot the paper on the suspected liquid agent.

c. Blot (do not rub) the M8 detector paper on the suspected liquid agent. Do not touch the liquid with your protective glove.

WARNING

Some decontaminants will give false positive results on the M8 detector paper. The M8 detector paper may indicate positive results if used in an area where decontaminants have been used.

---

d. Observe the M8 detector paper for a color change. Identify the contamination by comparing any color change on the M8 detector paper to the color chart on the inside front cover of the booklet.

(1) A yellow-gold color indicates the presence of a nerve (G) agent.

(2) A red-pink color indicates the presence of a blister (H) agent.

(3) A dark green color indicates the presence of a nerve (V) agent.

(4) Any other color or no color change indicates that the liquid cannot be identified using M8 detector paper.

e. Store the booklet of M8 detector paper.

f. Remain in MOPP4 even if the liquid cannot be identified. Use other types of chemical-agent detector kits to verify the test results.

g. Notify your supervisor of the test results.

Note: M8 detector paper reacts positively to petroleum products, ammonia, and decontaminating solution number 2 (DS2). M9 detector paper reacts positively to petroleum products, insecticides, and antifreeze. Because M9 detector paper only detects (but does not identify) chemical agents, verify all readings with M8 detector paper. If you observe a color change on M8 or M9 detector paper, assume it is a liquid chemical agent. When conducting agent tests at night, remove any colored lens because it may provide a false negative response. Confirm the presence of contamination by using all means of chemical-agent detection available in your area of operation, including a visual check of your surroundings. If you determine that your reading is a false positive, perform the following actions before giving the all clear signal:

1. Ensure that every attempt has been made to recheck the area.

2. Contact your higher headquarters (HQ) or the person in charge, and report the negative results.

3. Await further guidance. The higher HQ contacts all adjacent/attached units to check the status of contamination in their areas. If all units report the absence of contamination, the information is reported up the chain of command.

4. Annotate the above actions on DA Form 1594 Daily Staff Journal or Duty Officer's Log.

Evaluation Preparation:

> **CAUTION**
> Ensure that stimulants are placed on detector paper only and never on the protective clothing.

Setup: Provide the items listed in the task condition statement. Simulate an unknown liquid chemical agent by using expedient training aids (such as brake fluid, cleaning compound, gasoline, insect repellent, or antifreeze). Place drops of the simulated agent on M9 detector paper to obtain a reading. For M8 detector paper, place the simulated agent on nonporous material (such as an entrenching tool).

Brief Soldier: Tell the Soldier that he/she will be entering an area where chemical agents have been used. Tell him/her to attach M9 detector paper to his/her MOPP gear and equipment. Tell him/her that if you observe any acts that are unsafe or that could produce a false reading you will stop the test and he/she will be scored a NO GO.

Performance Measures	GO	NO GO
1 Detected chemical agents using M9 detector paper.	_____	_____
2 Detected chemical agents using M8 detector paper.	_____	_____

Evaluation Guidance: Score the Soldier GO if all performance measures are passed (P). Score the Soldier NO-GO if any performance measure is failed (F). If the Soldier fails any performance measure, show him how to do it correctly.

References:
Required: DA Form 1594, FM 3-11.3, FM 3-11.4, TM 3-6665-311-10, TM 3-6665-426-10
Related:

031-COM-1021

Mark CBRN-Contaminated Areas

Conditions: You are given a nuclear, biological, and chemical (NBC) marking set and Technical Manual (TM) 3-9905-001-10 or the M328 chemical, biological, radiological, and nuclear (CBRN) marking kit and in a tactical environment where CBRN weapons have been used. The contamination has been located and identified in an area. You are in the appropriate personal protective equipment (PPE). This task may be performed in mission-oriented protective posture (MOPP) level 4.

Standards: Mark the CBRN-contaminated area. Ensure that the required information is printed on the marker(s), and emplace the marker(s) according to the type of contamination. There is no change to standards if task is performed in MOPP level 4.

Special Condition: None

Special Standards: None

Special Equipment:

Cue: None

Note: If the M328 CBRN Marking Kit is available, proceed to step 2.

Performance Steps

1. Employ contamination markers using the NBC marking set.
   a. Emplace the RADIOLOGICAL markers.
      (1) Place markers at the location where a dose rate of 1 centigray per hour (cGyph) or more is measured.
      (2) Place markers so that the word "ATOM" faces away from the contamination.
      (3) Print the following information clearly on the front of the markers:
         (a) Dose rate in cGyph.
         (b) Date-time group (DTG) (specify local or Zulu) of the detonation. If the DTG is not known, print "unknown."
         (c) The DTG (specify local or Zulu) of the reading.
         (d) Go to steps d.
   b. Emplace the BIOLOGICAL markers.
      (1) Place markers at the location where contamination is detected.
      (2) Place markers so that the word "BIO" faces away from the contamination area.
      (3) Print the following information clearly on the front of the marker.
         (a) Name of agent, if known. If unknown, print "unknown."
         (b) DTG (specify local or Zulu) of detection.
      (4) Go to steps d.
   c. Emplace the CHEMICAL makers.
      (1) Place markers at the location where contamination is detected.

(2) Place markers so that the word "GAS" faces away from the contamination area.

(3) Print the following information clearly on the front of the marker:

(a) Name of agent, if known. If unknown, print "unknown."

(b) DTG (specify local or Zulu) of detection.

(4) Go to step d.

d. Position the markers so that the recorded information faces away from the area of contamination and place adjacent marking signs at intervals of 25 to 100 meters, depending on terrain.

e. If marking contamination in open terrain (e.g., desert, plains, rolling hills), raise markers to heights that permit approaching forces to view them at a distance up to 200 meters.

2. Employ contamination markers using the M328 CBRN Marking Kit.

a. Emplace the RADIOLOGICAL markers.

(1) Place markers at the location where a dose rate of 1 centigray per hour (cGyph) or more is measured.

(2) Place markers so that the word "ATOM" faces away from the contamination.

(3) Print the following information clearly on the front of the markers:

(a) Dose rate in cGyph.

(b) Date-time group (DTG) (specify local or Zulu) of reading.

(c) DTG of detonation, if known. If the DTG is not known, print "unknown."

(4) If beacons are required, proceed to step e; if not, proceed to step f.

b. Emplace the BIOLOGICAL markers.

(1) Place markers at the location where contamination is detected.

(2) Place markers so that the word "BIO" faces away from the contamination area.

(3) Print the following information clearly on the front of the markers:

(a) Name of agent, if known. If unknown, print "unknown."

(b) DTG (specify local or Zulu) of detection. If the DTG is not known, print "unknown."

(4) If beacons are required, proceed to step e; if not, proceed to step f.

c. Emplace the CHEMICAL markers.

(1) Place markers at the location where contamination is detected.

(2) Place markers so that the word "GAS" faces away from the contamination area.

(3) Print the following information clearly on the front of the marker.

(a) Name of agent, if known. If unknown, print "unknown."

(b) DTG (specify local or Zulu) of detection. If the DTG is not known, print "unknown."

(4) If beacons are required, proceed to step e; if not, proceed to steps f.

d. Emplace the toxic makers.

(1) Place markers at the location where contamination is detected.

(2) Place markers so that the word "TOXIC" faces away from the contamination area.

(3) Print the following information clearly on the front of the marker:

      (a) Name of agent, if known. If unknown, print "unknown."

      (b) DTG (specify local or Zulu) of detection. If the DTG is not known, print "unknown."

    (4) If beacons are required, proceed to step e; if not, proceed to steps f.

    e. Emplace beacons at approximately 300-meter intervals.

Note: Beacons are visible at night over ranges of up to 1,500+ meters. Beacons are supplied in visual and IR only types. Flexlight chemical lights are emplaced between beacons attached to the flag clips.

    f. Ensure that the recorded information on the markers faces away from the area contamination and place adjacent marking signs at intervals of 10 to 50 meters depending on terrain, approximately waist high. If beacons are used, the markers can be placed 10 to 100 meters apart.

    g. Ensure that when in open terrain all markers are at a height that permits approaching forces to view them at a distance up to 300 meters, approximately waist high.

Evaluation Preparation: *Setup*: Provide the Soldier with the items listed in the task condition statemSetup: Provide the Soldier with the items listed in the task condition statement. Use simulants to produce a contaminated environment for toxic and chemical or biological agents. For radiological contamination, tell the Soldier the type and amount of radiation present.

Brief Soldier: Tell the Soldier that the test will consist of ensuring that NBC markers are properly emplaced and that all required information is placed on the markers.

Performance Measures	GO	NO GO
1 Employed contamination markers using the NBC marking set.	_____	_____
2 Employed contamination markers using the M328 CBRN Marking Kit.	_____	_____

Evaluation Guidance: Score the Soldier GO if all performance measures are passed (P). Score the Soldier NO-GO if any performance measure is failed (F). If the Soldier fails any performance measure, show him how to do it correctly.

References:
Required: FM 3-11.3, ATP 3-11.37, TM 3-9905-001-10
Related:

031-COM-1018

React to Nuclear Hazard/Attack

Conditions: You are in a tactical situation or an area where nuclear weapons have been (or may have been) used. You are given loadbearing equipment (LBE), a piece of cloth or a protective mask, a brush or a broom, shielding material, Field Manual (FM) 3-11.3, and one of the following situations: 1. You see a brilliant flash of light. 2. You find a standard radiological contamination marker or an enemy marker. 3. You are told that fallout is in your area. 4. You receive instructions to respond to a nuclear attack. 5. You come across a suspected depleted-uranium (DU) hazard.

> **CAUTION**
> DO NOT USE MASKS WITH DAMAGED FILTERS BECAUSE CERTAIN MODELS CONTAIN HAZARDOUS MATERIALS. DO NOT CHANGE THE FILTER IN A CONTAMINATED ENVIRONMENT.

Standard: React to a nuclear hazard or attack with or without warning and without becoming a casualty. Identify radiological contamination markers with 100 percent accuracy, and notify your supervisor. Start the steps to decontaminate yourself within 1minute of finding radiological contamination. Decontaminate individual equipment after you completely decontaminate yourself.

Special Condition: None

Safety Level: Low

MOPP:

Performance Steps

1. React to a nuclear attack without warning.
   a. Close your eyes immediately.
   b. Drop to the ground in a prone position, facing the blast.

Note: If you are in the hatch of an armored vehicle, immediately drop down inside the vehicle.

   c. Keep your head and face down and your helmet on.
   d. Stay down until the blast wave passes and debris stops falling.
   e. Cover your mouth with a cloth or similar item to protect against inhalation of dust particles.
   f. Check for casualties and damaged equipment.
2. React to a nuclear attack with warning.
   a. Select the best available shelter.
      (1) Move into a fighting position, bunker, or ditch.

    (2) Take protective actions if you are inside a shelter.

    (3) Remain in place if you are in an armored vehicle.

  b. Protect your eyes.

  c. Minimize exposed skin areas.

  d. Cover your mouth with a cloth or similar item to protect against inhalation of dust particles.

3. React to a radiological contamination marker.

  a. Avoid the area, if possible.

  b. Cross the area quickly by the shortest route that exposes you to the least amount of radiation based on mission, enemy, terrain, troops, time available, and civilian considerations (METT-TC).

    (1) Request crossing instructions through the chain of command if you must cross.

    (2) Maximize the use of shielding.

    (3) Cover your mouth with a cloth or similar item to protect against inhalation of dust particles. A protective mask may be used if nothing else is available.

  c. Identify radiological contamination markers with 100 percent accuracy, and report the discovery of any markers identified to your supervisor.

4. Remove radiological contamination (including DU) from your clothing, equipment, and exposed skin.

  a. Shake or brush contaminated dust (all dust is considered to be radioactive) from your clothing, equipment, and exposed skin with a brush, a broom, or (if a brush or a broom is not available) your hands.

  b. Wash your body as soon as possible, giving special attention to hairy areas and underneath your fingernails.

  c. Conduct mission-oriented protection posture (MOPP) gear exchange if you are contaminated with wet radioactive contamination and were previously ordered to maintain a MOPP level.

(Asterisks indicate a leader performance step.)

	Performance Measures	GO	NO GO
1	React to a nuclear attack without warning.	_____	_____
2	React to a nuclear attack with warning.	_____	_____
3	React to a radiological contamination marker.	_____	_____
4	Remove radiological contamination (including DU)	_____	_____

Performance Measures	GO	NO GO

References:
Required: FM 3-11.5, FM 3-11.3
Related:

Environment: Environmental protection is not just the law but the right thing to do. It is a continual process and starts with deliberate planning. Always be alert to ways to protect our environment during training and missions. In doing so, you will contribute to the sustainment of our training resources while protecting people and the environment from harmful effects. Refer to FM 3-34.5 Environmental Considerations and GTA 05-08-002 ENVIRONMENTAL-RELATED RISK ASSESSMENT. Environmental protection is not just the law but the right thing to do. It is a continual process and starts with deliberate planning. Always be alert to ways to protect our environment during training and missions. In doing so you will contribute to the sustainment of our training resources while protecting people and the environment from harmful effects.

Safety: In a training environment, leaders must perform a risk assessment in accordance with ATP 5-19, Risk Management. Leaders will complete a DD Form 2977 DELIBERATE RISK ASSESSMENT WORKSHEET during the planning and completion of each task and sub-task by assessing mission, enemy, terrain and weather, troops and support available-time available and civil considerations, (METT-TC). Note: During MOPP training, leaders must ensure personnel are monitored for potential heat injury. Local policies and procedures must be followed during times of increased heat category in order to avoid heat related injury. Consider the MOPP work/rest cycles and water replacement guidelines IAW FM 3-11.4, Multiservice Tactics, Techniques, and Procedures for Nuclear, Biological, and Chemical (NBC) Protection, FM 3-11.5, Multiservice Tactics, Techniques, and Procedures for Chemical, Biological, Radiological, and Nuclear Decontamination. Everyone is responsible for safety. A thorough risk assessment must be completed prior to every mission or operation.

Supporting tasks:
031-503-1028 Operate the AN/PDR-77 Radiac Set

031-COM-1042

Protect Yourself from CBRN Injury/Contamination when Changing MOPP using the JSLIST Chemical Protective Ensemble.

Conditions: You are in mission-oriented protective posture (MOPP) 4 with individual gear and equipment. Your MOPP gear is contaminated. Your buddy is in MOPP4 with individual gear and equipment, available to assist you with MOPP gear exchange. You have an uncontaminated set of MOPP gear for yourself and your buddy, a personal decontamination kit, a M295 individual equipment decontamination kit (IEDK), FM 3-11.4 and FM 3-11.5, cutting tools, M100 sorbent decontamination system (SDS), an improved chemical-agent monitor (ICAM), three 3-gallon pails, sponges, paper towels, soap, and water. This task will be performed in MOPP4.

Standards: Protect yourself from chemical, biological, radiological, and nuclear (CBRN) injury/contamination when changing MOPP using the JSLIST, by (1) decontaminating individual gear and equipment (without spreading contamination onto your skin or undergarments), (2) setting uncontaminated gear aside on an uncontaminated surface, (3) changing overgarments, overboots, and gloves (without spreading contamination to the uncontaminated set of MOPP gear), and (4) changing MOPP gear (without you or your buddy becoming a casualty).

Special Condition: None

Safety Level: Low

Cue: You are in MOPP Level 4 with individual gear and equipment. Your MOPP gear is contaminated. Your buddy is in MOPP 4 with individual gear and equipment, available to assist you with MOPP gear exchange. You have an uncontaminated set of chemical MOPP gear for yourself and your buddy, personal decontamination kit, an M295 individual equipment decontamination kit (IEDK), M100 sorbent decontamination system (SDS), FM 3-11.4 and FM 3-11.5, cutting tools, an improved chemical-agent monitor (ICAM), three 3-gallon pails, sponges, paper towels, soap, and water.

Note: Both Soldiers will perform steps 1 and 2 at the same time. If, during the technique, it is suspected that contamination has spread onto their skin or undergarments, both Soldiers will decontaminate immediately with the available IEDK, and then proceed with the MOPP gear exchange.

WARNING:
THE JOINT SERVICE, LIGHTWEIGHT, INTEGRATED SUIT
TECHNOLOGY (JSLIST) IS DESIGNED TO PROTECT SOLDIERS FROM
TRADITIONAL NUCLEAR, BIOLOGICAL, AND CHEMICAL THREATS
NOT THE FULL SPECTRUM OF INDUSTRIAL CHEMICAL HAZARDS.

CAUTION:
WHEN REMOVING THE IMPROVED OUTER TACTICAL VEST (IOTV)
OVER THE SHOULDER TAKE CARE NOT TO SNAG THE FILTER
CANISTER AND BREAK THE SEAL OF YOUR PROTECTIVE MASK

Performance Steps

1. Decontaminate individual gear for chemical or biological contamination without assistance.

    a. Remove and discard the chemical protective helmet cover.

    b. Rub the M295 or M100 SDS into the material.

    c. Shake the excess off gently.

    d. Set the gear aside on an uncontaminated surface (such as a poncho, a canvas, or similar material).

2. Decontaminate individual gear for radiological contamination without assistance.

    a. Brush, wipe, or shake off the dust for radiological contamination from the individual gear.

    b. Wash the equipment with warm, soapy water.

    c. Set the equipment aside to dry on an uncontaminated surface.

3. Prepare for decontamination.

    a. Buddy: Remove the M9 paper; untie the bow in the coat retention cord, if tied; unfasten the webbing strip snap at the bottom front of the coat; and release the waist coat retention cord loop.

    b. Buddy: Loosen the bottom of the coat by pulling the material away from the body.

    c. Feel for the suspender snap couplers on the outside of the coat, and release the snap couplers.

    d. Unfasten the hook-and-pile fasteners at the wrist and ankles, and refasten them loosely.

    e. Unfasten the two strap-and-buckle fasteners on the multipurpose overboots (MULOs) and unfasten or cut the fasteners on the black vinyl overboots (BVOs), or untie/cut the laces on the chemical-protective overboots.

4. Decontaminate the mask and hood.

    a. Chemical or biological contamination.

        (1) The buddy uses M295 to decontaminate the exposed parts of the mask, instructing the Soldier to put two fingers on the voicemitter to avoid breaking the seal.

        (2) The buddy starts at the eye lens outserts and wipes all exposed parts of the mask.

        (3) The buddy wipes the front edge of the hood, including the barrel locks and fasteners under your chin.

(4) The buddy decontaminates his/her gloves in preparation to release the hood seal.

b. Radiological contamination.

(1) The buddy wipes your mask with warm, soapy water.

Note: Cool, soapy water is not as effective for removing contamination, but it can be used if the material is scrubbed longer.

(2) The buddy rinses your mask with a sponge dipped in clean water.

(3) The buddy dries your mask with paper towels or rags.

(4) The buddy decontaminates his/her gloves in preparation to release the hood seal.

5. Doff the chemical-protective coat.

a. The buddy unties the draw cord, if tied, presses the barrel lock release, and unsnaps the barrel locks.

Note: If the buddy has difficulty grasping the barrel locks, use the draw cord to pull the locks away from the mask, allowing the buddy to grasp and unfasten the locks without touching the hood's interior.

b. The buddy unfastens the front closure flap and slides the fastener from the chin to the bottom of the coat.

c. The buddy instructs the Soldier to turn around, grasps the hood, and rolls it inside out (pulling the hood off the Soldier's head).

d. The buddy grasps the coat at the shoulders and instructs the Soldier to make a fist to prevent the chemical protective gloves from coming off.

e. The buddy pulls the coat down and away from the Soldier, ensuring that the black part of the coat is not touched.

Note: If there is difficulty removing the coat in this manner, pull one arm off at a time.

CAUTION:
BOTH SOLDIERS MUST TAKE CARE TO AVOID CONTAMINATING THE INSIDE SURFACE OF THE COAT BECAUSE IT WILL BE USED LATER AS AN UNCONTAMINATED SURFACE TO STAND ON DURING THE DONNING PROCEDURES.

---

f. The buddy lays the coat on the ground, black side up.

6. Doff the chemical-protective trousers.

a. Unfasten the hook-and-pile fastener tapes at the waistband, unfasten the two front closure snaps, and open the fly slide fastener on the front of the trousers.

b. Buddy: Grasp the trousers at the hips, and pull them down to the knees.

c. Buddy: Have the Soldier lift one leg (with the foot pointed down). With your hand on each side, pull the trousers in an alternating motion until the Soldier can step out of the trouser leg. Repeat the process for the other leg.

> **CAUTION:**
> BOTH SOLDIERS MUST TAKE CARE TO AVOID CONTAMINATING
> THEIR CLOTHING AND SKIN.

    d. Discard the trousers away from the clean area.

   7. Doff the chemical-protective overboots.

    a. Buddy: Remove the chemical-protective overboots while the Soldier is standing with his/her arms up, shoulder high, to avoid contaminating the clothing or skin.

Note: The Soldier may put a hand on the buddy for balance, but he/she must then decontaminate the gloves.

    b. Stand next to the coat spread on the ground.

    c. Remove one overboot by stepping on a heel with one foot while pulling the other foot upward.

    d. Buddy: Pull off the Soldier's overboots, one foot at a time.

    e. Step on the coat that is spread on the ground as each foot is withdrawn from the overboot.

> **CAUTION:**
> THE BUDDY MUST TAKE CARE TO AVOID TOUCHING THE
> SOLDIER'S COMBAT BOOTS. THE SOLDIER MUST TAKE CARE TO
> AVOID LETTING THE COMBAT BOOTS TOUCH THE GROUND.

    f. Discard the overboots away from the clean area.

   8. Doff the chemical-protective gloves and liners.

    a. Hold the fingertips of the gloves, and partially slide your hand out.

    b. Hold your arms away from your body when both hands are free. Let the gloves drop off and away from the black side of the coat.

    c. Remove the protective glove inserts.

> **CAUTION:**
> BOTH SOLDIERS MUST TAKE CARE TO AVOID LETTING THE
> GLOVES MAKE CONTACT WITH THE COAT THAT IS SPREAD ON
> THE GROUND.

    d. Buddy: Discard the Soldier's chemical-protective gloves and inserts away from the clean area.

   9. Don the chemical-protective trousers.

    a. Buddy: Open the package containing the new trousers, but do not touch the inside of the package.

    b. Stand on the uncontaminated surface. Reach into the package, and remove the trousers without touching the outside of the package.

    c. Put on the trousers, close the slide fastener, and fasten the two fly opening snaps. Pull the suspenders over your shoulders, and fasten the snap couplers.

    d. Adjust the length of the suspenders to ensure that you have a comfortable fit in the inseam.

> **CAUTION:**
> THE SOLDIER MUST TAKE CARE TO ENSURE THAT THE TROUSERS TOUCH ONLY THE UNCONTAMINATED SURFACE.

    e. Adjust the hook-and-pile fasteners at the waistband for a snug fit.

10. Don the chemical-protective coat.

    a. Buddy: Open the package containing the new coat, and have the Soldier reach in and remove the coat. Be careful not to touch the outside of the package.

    b. Don the coat. Close the slide fastener up as far as the chest, and secure the front closure hook-and-pile fastener tape on the front flap as far as the chest.

    c. Pull the bottom of the coat down over the trousers. Grasp the loop on the back of the overgarment coat, pull the loop out and away from the overgarment coat, and bring the loop forward between your legs, pulling on the loop so that the bottom of the coat fits snugly over the trousers.

> **CAUTION:**
> THE SOLDIER MUST TAKE CARE TO ENSURE THAT HIS/HER BODY AND CLOTHING TOUCH ONLY THE INNER SURFACE OF THE COAT.

    d. Place the loop over the webbing strip on the front of the coat, and fasten the strap on the webbing strip to keep the loop in place. Adjust the retention cord on the coat, if necessary. Tie any excessive cord in a bow.

11. Don the chemical-protective over-boots.

    a. Buddy: Open the package containing the new over-boots, and have the Soldier remove the over-boots, being careful not to touch the outside of the package.

    b. Don the over-boots over the combat boots. Adjust and secure the strap-and-buckle fasteners. Pull the trouser legs over the over-boots, and secure the two hook-and-pile fastener tapes on each ankle to fit snugly around the over-boot.

12. Don the chemical-protective hood.

    a. Put the hood on. Close the front slide fastener on the coat completely, and secure the hook-and-pile fastener tape on the front flap as far as the top of the slide fastener.

> **WARNING:**
> THE BARREL LOCK RELEASE BUTTON MUST FACE AWAY FROM THE USER WHEN WORN TO PREVENT THE LOCK FROM UNFASTENING AND POSSIBLY EXPOSING THE USER TO CONTAMINATION.

    b. Place the edge of the hood around the edge of the mask, and secure the hook-and-pile fastener tape on the hood.

    c. Pull the draw cord tight around the edge of the mask, snap the barrel locks together, squeeze both ends of the lock while pulling the draw cord, and slide the barrel lock up under the chin to keep the cord in place.

d. Buddy: Inspect the hood and mask to ensure that the hood is positioned properly, the skin is not exposed, and any excessive draw cord is tied in a bow, without touching the Soldier.

e. Adjust as directed.

Note: If buddy assistance is required for proper adjustment, the buddy decontaminates his/her gloves before touching the Soldier's hood or mask.

13. Don the chemical-protective gloves and liners.

a. Buddy: Open the package containing the new chemical-protective gloves and liners. The Soldier removes the gloves and liners, being careful not to touch the outside of the package.

b. Don the liners and gloves, pull the cuffs of the coat over the chemical-protective gloves, and fasten the hook-and pile fasteners on each coat sleeve.

c. Put on the M9 chemical-agent detection paper as required by the standing operating procedure (SOP).

14. Reverse roles with the buddy, and repeat steps 3 through 13.

CAUTION:
WHEN DONNING THE IOTV OVER THE SHOULDER TAKE CARE NOT TO SNAG THE FILTER CANISTER AND BREAK THE SEAL OF YOUR PROTECTIVE MASK.

---

15. Secure individual gear.

a. Place a new chemical-protective helmet cover on the helmet, if a personnel armor system, ground troop (PASGT) helmet is used.

b. Use the buddy system to check the fit of the gear.

Evaluation Guidance: Brief Soldier: Identify buddy pairs, designating the initial task performer and the buddy. Provide each Soldier with one of the following three scenarios: (1) The Soldier has been exposed to chemical or biological contamination (Steps 2 and 4b are omitted), (2) The Soldier has been exposed to radiological contamination (Steps 1 and 4a are omitted), or (3) The Soldier has been exposed to radiological and chemical and/or biological contamination (all steps must be performed).

Evaluation Preparation: Setup: Provide the Soldier with the items listed in the task conditions statement. Evaluate this task during field exercises or normal training sessions. Soldiers must be in MOPP4.

Performance Measures	GO	NO GO
1 Decontaminated individual gear for chemical or biological contamination without assistance.	____	____

Performance Measures		GO	NO GO
2	Decontaminated individual gear for radiological contamination wothout assistance	_____	_____
3	Prepare for decontamination.	_____	_____
4	Decontaminated the mask and hood.	_____	_____
5	Doffed the chemical-protective coat.	_____	_____
6	Doffed the chemical-protective trousers.	_____	_____
7	Doffed the chemical-protective overboots.	_____	_____
8	Doffed the chemical-protective gloves and liners.	_____	_____
9	Donnned the chemical-protective trousers.	_____	_____
10	Donnned the chemical-protective coat.	_____	_____
11	Donnned the chemical-protective overboots.	_____	_____
12	Donnned the chemical-protective hood.	_____	_____
13	Donnned the chemical-protective gloves and liners.	_____	_____
14	Reversed roles with the buddy, and repeated steps 3 through 13.	_____	_____
15	Secured individual gear	_____	_____

References:

Performance Measures	GO	NO GO

Required: FM 3-11.3, FM 3-11.4, FM 3-11.5, FM 4-25.11, TC 3-11-55, TM 10-8415-220-10, TM 3-4230-235-10, TM 3-4230-236-10, TM 3-4240-346-10
Related: ATP 5-19, GTA 05-08-002

Environment: Environmental protection is not just the law but the right thing to do. It is a continual process and starts with deliberate planning. Always be alert to ways to protect our environment during training and missions. In doing so, you will contribute to the sustainment of our training resources while protecting people and the environment from harmful effects. Refer to FM 3-34.5 Environmental Considerations and GTA 05-08-002 ENVIRONMENTAL-RELATED RISK ASSESSMENT. Environmental protection is not just the law but the right thing to do. It is a continual process and starts with deliberate planning. Always be alert to ways to protect our environment during training and missions. In doing so you will contribute to the sustainment of our training resources while protecting people and the environment from harmful effects.

Safety: In a training environment, leaders must perform a risk assessment in accordance with ATP 5-19, Risk Management. Leaders will complete a DD Form 2977 DELIBERATE RISK ASSESSMENT WORKSHEET during the planning and completion of each task and sub-task by assessing mission, enemy, terrain and weather, troops and support available-time available and civil considerations, (METT-TC). Note: During MOPP training, leaders must ensure personnel are monitored for potential heat injury. Local policies and procedures must be followed during times of increased heat category in order to avoid heat related injury. Consider the MOPP work/rest cycles and water replacement guidelines IAW FM 3-11.4, Multiservice Tactics, Techniques, and Procedures for Nuclear, Biological, and Chemical (NBC) Protection, FM 3-11.5, Multiservice Tactics, Techniques, and Procedures for Chemical, Biological, Radiological, and Nuclear Decontamination. Everyone is responsible for safety. A thorough risk assessment must be completed prior to every mission or operation.

081-COM-1044

Perform First Aid for Nerve Agent Injury

Condition: You and your unit are in an area where there is a threat of chemical attack. You are wearing protective overgarments and/or mask, or they are immediately available. There are casualties with possible nerve agent injuries. You will need chemical protective gloves, overgarments, overboots, protective mask and hood, mask carrier, and nerve agent antidote auto-injectors. The casualty has three sets of MARK I nerve agent antidote auto-injectors or three antidote treatment nerve agent auto-injector (ATNAAs) or one convulsant antidote for nerve agents (CANA) auto-injector. Some iterations of this task should be performed in MOPP.

Standard: Administer the antidote correctly to yourself, or administer three sets of MARK I nerve agent antidote auto-injectors or three ATNAAs followed by the CANA to a buddy following the correct sequence.

Special Condition: None

Safety Level: Low

MOPP: Sometimes

Cue: None

Remarks: None

Performance Steps

Cue: Soldiers have come under possible chemical attack.
1. React to the chemical hazard.
   a. Stop breathing immediately and close your eyes.
   b. Don your protective mask. (See task 031-COM-1035).
Note: Do NOT put on additional protective clothing at this time.
   c. Give the alarm.
Note: Information on this step is provided in task 031-COM-1019.
2. Identify signs and symptoms of nerve agent poisoning.
   a. Mild nerve agent poisoning.
Note: For signs and symptoms of mild nerve agent poisoning, first aid is considered to be self-aid.
    (1) Unexplained runny nose.
    (2) Unexplained sudden headache.
    (3) Sudden drooling.
    (4) Tightness in the chest or difficulty breathing.
    (5) Difficulty seeing (dimness of vision or miosis).
    (6) Localized sweating and muscular twitching in the area of contaminated skin.
    (7) Stomach cramps.
    (8) Nausea.
Note: For the above signs and symptoms, first aid is considered to be self-aid.
   b. Severe nerve agent poisoning.
Note: For signs and symptoms of severe nerve agent poisoning, first aid is considered to be buddy-aid.
    (1) Strange or confused behavior.
    (2) Wheezing, difficulty breathing and coughing.
    (3) Severely pinpointed pupils.
    (4) Red eyes with tearing.
    (5) Vomiting.
    (6) Severe muscular twitching.
    (7) Involuntary urination and defecation.
    (8) Convulsions.

(9) Unconsciousness and/or respiratory failure.

    c. Localized sweating and muscular twitching in the area of contaminated skin.

Cue: Signs and symptoms of nerve agent poisoning have been identified.

    3. Administer self-aid for mild nerve agent poisoning.

Note: Only administer one MARK I or ATNAA as self-aid. Do not self-administer the CANA.

    a. MARK I.

        (1) Obtain one MARK I auto-injector.

        (2) Locate injection site (outer thigh muscle, about a hand's width below the hip joint and above the knee) and ensure that it is clear of objects that will interfere with injection.

Note: If the individual is thinly built, injection should be given into the upper outer quadrant of the buttock.

        (3) With non-dominate hand, hold the set of injectors by the plastic clip at eye level with the large injector on top.

        (4) With other hand, grasp the atropine (smaller) injector without covering or holding the needle (green) end, and pull it out of the clip, forming a fist around the auto-injector, with the green end extending just past the little finger of your fist.

Note: If the injection is accidentally given in the hand, another small injector must be obtained and the injection given in the proper site.

CAUTION

When injecting antidote in the buttock, be very careful to inject only in the upper, outer quarter of the buttock to avoid hitting the major nerve that crosses the buttocks. Hitting the nerve may cause paralysis.

---

        (5) Place the needle end of the injector against chosen injection site and apply firm pressure until needle activates into muscle.

Note: A jabbing motion is not necessary to trigger the activating mechanism.

        (6) Massage the injection site, mission permitting.

        (7) Remove the injector from your muscle and carefully place this used injector between two fingers of the hand holding the plastic clip.

        (8) Pull the 2 PAM CI (larger) injector out of the clip and form a fist around the auto-injector with the needle (black) end extending beyond the little finger. Drop the clip to the ground.

        (9) Place the needle end of the injector against the injection site.

        (10) Secure the used injectors.

        (11) Bend the needles of all used injectors by pressing on a hard surface to form a hook.

        (12) Attach all used injectors to blouse pocket flap or Joint Service Lightweight Integrated Suit Technology (JSLIST).

> **WARNING**
>
> Do NOT give yourself additional injections. If you are able to walk without assistance and know who you are and where you are, you will NOT need the second set of injections. If you continue to have symptoms of nerve agent poisoning, seek someone else (a buddy) to check your symptoms and administer the additional sets of injections, if required.

      (13) Massage the injection site, mission permitting.

    b. ATNAA.

      (1) Obtain one ATNAA auto-injector.

      (2) Locate injection site (outer thigh muscle, about a hand's width below the hip joint and above the knee) and ensure that it is clear of objects that will interfere with the injection.

Note: If the individual is thinly built, injection should be given into the upper outer quadrant of the buttock.

> **CAUTION**
>
> Do NOT cover or hold the needle end with your hand, thumb, or fingers. You may accidentally inject yourself.

      (3) With your dominant hand, hold the ATNAA in your closed fist with the needle (green) end extending beyond the little finger in front of you at eye level.

      (4) Pull off the safety cap from the bottom of the injector with a smooth motion using non-dominate hand, and drop it to the ground.

> **CAUTION**
>
> When injecting antidote in the buttock, be very careful to inject only into the upper, outer quarter of the buttock to avoid hitting the major nerve that crosses the buttocks. Hitting the nerve may cause paralysis.

      (5) Place the needle end of the injector against chosen injection site and apply firm, even pressure until needle activates into muscle.

Note: A jabbing motion is not necessary to trigger the activating mechanism.

      (6) Hold the injector firmly in place for at least 10 seconds.

      (7) Remove the injector from your muscle.

      (8) Secure the used injector.

      (9) Bend the needles of all used injectors by pressing on a hard surface to form a hook.

      (10) Attach all used injectors to blouse pocket flap or JSLIST.

(11) Massage the injection site, mission permitting.

Cue: The casualty is masked.

4. Administer buddy-aid for severe nerve agent poisoning.

a. Mask the casualty if necessary.

(1) Place the mask on the casualty.

(2) If the casualty can follow directions, have him clear the mask.

(3) Check for a complete mask seal by covering the inlet valves of the mask.

(4) Pull the protective hood over the head, neck and shoulders of the casualty.

(5) Position the casualty on the right side (recovery position) with the head slanted down so that the casualty will not roll back over.

b. MARK I.

Note: Before initiating buddy-aid, determine if one ATNAA or one set of MARK I auto-injectors have already been used. No more than three sets (total) of the antidote are to be administered.

(1) Position yourself near the casualty's thigh.

(2) Obtain casualty's three or remaining MARK I auto-injectors.

Note: Be sure to use the casualty's auto-injector and not your own.

(3) Using the same method as in self-aid, administer up to, but no more than three doses of the MARK I nerve agent antidote.

Note: If casualty's condition improves (regains consciousness, becomes coherent, able to stand or walk) after the first or second dose, do not administer the third dose but monitor until help arrives or he is evacuated to higher care.

(4) Bend the needles of all used injectors by pressing on a hard surface to form a hook.

(5) Attach all used injectors to blouse pocket flap or JSLIST.

c. ATNAA.

(1) Position yourself near the casualty's thigh.

(2) Obtain casualty's three or remaining ATNAA auto-injectors.

Note: Be sure to use the casualty's own auto-injectors, and not your own.

(3) Using the same method as in self-aid, administer up to, but no more than three doses of the ATNAA nerve agent antidote.

Note: If casualty's condition improves (regains consciousness, become coherent, able to stand or walk) after the first or second dose, do not administer the remaining dose(s), but monitor until medical help arrives or he is evacuated to higher care.

(4) Bend the needles of all used injectors by pressing on a hard surface to form a hook.

(5) Attach all used injectors to blouse pocket flap or JSLIST.

d. CANA.

Note: Buddy-aid also includes administering the CANA with the third MARK I or ATNAA to prevent convulsions.

CAUTION
Squat, do not kneel, when masking the casualty or administering the nerve agent antidotes to the casualty.

(1) Position yourself near the casualty's thigh.

(2) Obtain one CANA auto-injector.

(3) Locate injection site (outer thigh muscle, about a hand's width below the hip joint and above the knee) and ensure that it is clear of objects that will interfere with the injection.

Note: If the individual is thinly built, injection should be given into the upper outer quadrant of the buttock.

CAUTION
Do NOT cover or hold the needle end with your hand, thumb, or fingers. You may accidentally inject yourself.

(4) With your dominant hand, hold the CANA in your closed fist with the needle end extending beyond the little finger in front of you at eye level.

(5) Pull off the safety cap from the bottom of the injector with a smooth motion using non-dominant hand, and drop it to the ground.

CAUTION
When injecting antidote in the buttock, be very careful to inject only into the upper, outer quarter of the buttock to avoid hitting the major nerve that crosses the buttocks. Hitting the nerve may cause paralysis.

(6) Place the needle end of the injector against chosen injection site and apply firm, even pressure until needle activates into muscle.

(7) Hold the injector firmly in place for at least 10 seconds.

(8) Remove the injector from casualty's muscle.

(9) Secure the used injector.

5. Decontaminate skin, if necessary.

Note: Information on this step is provided in task 031-COM-1013.

6. Put on remaining protective clothing.

Note: Information on this step is covered in task 031-COM-1040.

7. Seek medical aid.

Evaluation Guidance: For step 2, tell the Soldier to state, in any order, the mild symptoms of nerve agent poisoning. The Soldier must state seven of the eight symptoms to be scored GO. Tell the Soldier that he has mild symptoms and must take appropriate action. After Soldier completes step 3, ask what should be done next. Then ask what he should do after putting on all protective clothing. Score steps 5 through 7 based on the Soldier's responses. For step 4, tell the Soldier to state, in any order, the severe symptoms of nerve agent poisoning. The Soldier must state eight of the nine symptoms to be scored GO. Tell the Soldier to treat the casualty for nerve agent poisoning.

Evaluation Preparation: You must use nerve agent antidote injection training aids to train and evaluate this task. Actual auto-injectors will not be used. For self-aid, have the Soldier dress in MOPP2. Have the Soldier wear a mask carrier containing a mask and the training nerve agent auto-injectors. For buddy aid, have the Soldier being tested and the casualty dress in MOPP2. Have the casualty lie on the ground wearing the mask carrier containing a mask and the training nerve agent auto-injectors.

Performance Measures	GO	NO GO
1 Reacted to the chemical hazard.	____	____
2 Identified signs and symptoms of nerve agent poisoning.	____	____
3 Administered self-aid for mild nerve agent poisoning.	____	____
4 Administered buddy-aid for severe nerve agent poisoning.	____	____
5 Decontaminated skin, if necessary.	____	____
6 Put on remaining protective clothing.	____	____
7 Sought medical aid.	____	____

References:
Required: ATP 4-02.285, FM 4-25.11
Related:

Environment: Environmental protection is not just the law but the right thing to do. It is a continual process and starts with deliberate planning. Always be alert to ways to protect our environment during training and missions. In doing so, you will contribute to the sustainment of our training resources while protecting people and the environment from harmful effects. Refer to FM 3-34.5 Environmental

Considerations and GTA 05-08-002 ENVIRONMENTAL-RELATED RISK ASSESSMENT. Environmental protection is not just the law but the right thing to do. It is a continual process and starts with deliberate planning. Always be alert to ways to protect our environment during training and missions. In doing so, you will contribute to the sustainment of our training resources while protecting people and the environment from harmful effects. Refer to FM 3-34.5 Environmental Considerations and GTA 05-08-002 ENVIRONMENTAL-RELATED RISK ASSESSMENT.

Safety: In a training environment, leaders must perform a risk assessment in accordance with ATP 5-19, Risk Management. Leaders will complete a DD Form 2977 DELIBERATE RISK ASSESSMENT WORKSHEET during the planning and completion of each task and sub-task by assessing mission, enemy, terrain and weather, troops and support available-time available and civil considerations, (METT-TC). Note: During MOPP training, leaders must ensure personnel are monitored for potential heat injury. Local policies and procedures must be followed during times of increased heat category in order to avoid heat related injury. Consider the MOPP work/rest cycles and water replacement guidelines IAW FM 3-11.4, Multiservice Tactics, Techniques, and Procedures for Nuclear, Biological, and Chemical (NBC) Protection, FM 3-11.5, Multiservice Tactics, Techniques, and Procedures for Chemical, Biological, Radiological, and Nuclear Decontamination.

081-COM-1001

Evaluate a Casualty (Tactical Combat Casualty Care)

Conditions: While in a tactical area of operations, you encounter a combat casualty. Your unit may be under fire.

Some iterations of this task should be performed in MOPP.

Standards: Evaluate the casualty following the correct sequence. Identify and treat all life-threatening conditions and other serious wounds.

Special Condition: None

Special Standards: None

Safety Level: Low

MOPP: Sometimes

Special Equipment:

Cue: None

Note: None

WARNING

If a broken neck or back is suspected, do not move the casualty unless to save his/her life.

---

1. Perform care under fire.
    a. Return fire as directed or required before providing medical treatment.
    b. Determine if the casualty is alive or dead.

Note: In combat, the most likely threat to the casualty's life is from bleeding. Attempts to check for airway and breathing will expose the rescuer to enemy fire. Do not attempt to provide first aid if your own life is in imminent danger. In a combat situation, if you find a casualty with no signs of life--no pulse, no breathing--do NOT attempt to restore the airway. Do NOT continue first aid measures.

    c. Provide care to the live casualty. Direct the casualty to return fire, move to cover, and administer self-aid (stop bleeding), if possible.

Note: Reducing or eliminating enemy fire may be more important to the casualty's survival than the treatment you can provide.

If the casualty is unable to move and you are unable to move the casualty to cover and the casualty is still under direct enemy fire, have the casualty "play dead."

Cue: Enemy fire has been suppressed

    d. In a battle-buddy team, approach the casualty (use smoke or other concealment if available using the most direct route possible.
    e. Administer life-saving hemorrhage control.
        (1) Determine the relative threat of enemy fire versus the risk of the casualty bleeding to death.
        (2) If the casualty has severe, life-threatening bleeding from an extremity or has an amputation of an extremity, administer life-saving hemorrhage control by applying a tourniquet from the casualty's IFAK before moving the casualty. (See task 081-COM-1032.)

Note: The only treatment that should be given at the point of injury is a tourniquet to control life-threatening extremity bleeding.

    f. Move the casualty, his weapon, and mission-essential equipment when the tactical situation permits.
    g. Recheck bleeding control measures (tourniquet) as soon as behind cover and not under enemy fire.

Cue: You are now behind cover and are not under hostile fire.

2. Perform tactical field care.

Note: When evaluating and/or treating a casualty, seek medical aid as soon as possible. Do NOT stop treatment. If the situation allows, send another person to find medical aid.

    a. Form a general impression of the casualty as you approach (extent of injuries, chance of survival).

Note: If a casualty is being burned, take steps to remove the casualty from the source of the burns before continuing evaluation and treatment. (See task 081-COM-1007.)

    (1) Ask in a loud, but calm, voice: "Are you okay?" Gently shake or tap the casualty on the shoulder.

    (2) Determine the level of consciousness by using AVPU: A = Alert; V = responds to Voice; P = responds to Pain; U = Unresponsive.

Note: To check a casualty's response to pain, rub the breastbone briskly with a knuckle or squeeze the first or second toe over the toenail. If casualty is wearing IBA, pinch his nose or his earlobe for responsiveness.

    (3) If the casualty is conscious, ask where his body feels different than usual, or where it hurts.

Note: If the casualty is conscious but is choking and cannot talk, stop the evaluation and begin treatment. (See task 081-COM-1003.)

    c. Identify and control bleeding.

      (1) Check for bleeding.

        (a) Reassess any tourniquets placed during the care under fire phase to ensure they are still effective.

        (b) Perform a blood sweep of the extremities, neck, axillary, inguinal and extremity areas. Exposure is only necessary if bleeding is detected.

        (1) Place your hands behind the casualty's neck and pass them upward toward the top of the head. Note: whether there is blood or brain tissue on your hands from the casualty's wounds.

        (2) Place your hands behind the casualty's shoulders and pass them downward behind the back, the thighs, and the legs. Note whether there is blood on your hands from the casualty's wounds.

Note: If life-threatening bleeding is present, stop the evaluation and control the bleeding. (See task 081- COM-1032).

      (3) Once bleeding has been controlled, continue to step 2d.

    d. Position the casualty and open the airway. (See task 081-COM-1023.)

    e. Assess for breathing and chest injuries.

      (1) Expose the chest and check for equal rise and fall and for any wounds.

      (2) Look, listen, and feel for respiration. (See task 081-COM-1023.)

Note: If the casualty is breathing, insert a nasopharyngeal airway (see task 081-COM-1023.) and place the casualty in the recovery position.

Only in the case of non-traumatic injuries such as hypothermia, near drowning, or electrocution should CPR be considered when in a tactical environment prior to the CASEVAC phase.

      (3) If in a non-tactical environment, begin rescue breathing as necessary to restore breathing and/or pulse (See tasks 081-COM-1023 and 081-COM-0046.).

        (a) If the casualty has a penetrating chest wound and is breathing or attempting to breathe, stop the evaluation to apply an occlusive dressing (See task 081-COM-1026.).

        (b) Position or transport with the affected side down, if possible.

        (c) Check for an exit wound. If found, apply an occlusive dressing.

f. Dress all non-life threatening injuries and any bleeding that has not been addressed earlier with appropriate dressings. (See task 081-COM-1032.)

3. Determine the need to evacuate the casualty and supply information for lines 3-5 of the 9-Line MEDEVAC request to your tactical leader. (See task 081-COM-0101.)

4. Check the casualty for burns.

a. Look carefully for reddened, blistered, or charred skin. Also check for singed clothes.

b. If burns are found, stop the evaluation and begin treatment. (See task 081-COM-1007.)

5. Administer pain medications and antibiotics (the casualty's combat pill pack) if available.

Note: Each Soldier will be issued a combat pill pack before deploying on tactical missions.

6. Document the injuries and the treatment given on the casualty's own Tactical Combat Casualty Care Card (found in IFAK), if applicable.

Note: The FMC is usually initiated by the combat medic. However, a certified combat lifesaver can initiate the FMC if a combat medic is not available or if the combat medic directs the combat lifesaver to initiate the card. A pad of FMCs is part of the combat lifesaver medical equipment set.

7. Transport the casualty to the evacuation site. (See task 081-COM-1046.)

8. Monitor the patient for shock and treat as appropriate. (See task 081-COM-1005.) Continually reassess casualty until a medical person arrives or the patient arrives at a military treatment facility (MTF).

Evaluation Preparation:

Setup: Prepare a "casualty" for the Soldier to evaluate in step 2 by simulating one or more wounds or conditions. Simulate the wounds using a war wounds moulage set, casualty simulation kit, or other available materials. You can coach a "conscious casualty" on how to respond to the Soldier's questions about location of pain or other symptoms of injury. However, you will have to cue the Soldier during evaluation of an "unconscious casualty" as to whether the casualty is breathing and describe the signs or conditions, as the Soldier is making the checks.

Brief Soldier:

Performance Measures	GO	NO GO
1 Performed care under fire.	_____	_____

Performance Measures	GO	NO GO
2 Performed tactical field care.	_____	_____
3 Determined need to evacuate and reported information to tactical leader.	_____	_____
4 Checked the casualty for burns.	_____	_____
5 Administered pain medication and antibiotics (if applicable).	_____	_____
6 Documented injuries found on the Tactical Combat Casualty Card.	_____	_____
7 Transported the casualty to evacuation site.	_____	_____
8 Monitored the patient for signs and symptoms of shock.	_____	_____

Evaluation Guidance: Tell the Soldier to do, in order, all necessary steps of Tactical Combat Casualty Care and treat all wounds and/or conditions identified appropriately. Tell the Soldier that he/she will not perform first aid but will tell you what first aid action (give mouth-to-mouth resuscitation, bandage the wound, and so forth) he/she would take. After he/she has completed the checks ask him/her what else should be done. To test step 8, ask the Soldier what should be? While evacuating an unconscious casualty. Tell the Soldier to do, in order, all necessary steps of Tactical Combat Casualty Care and treat all wounds and/or conditions identified appropriately. Tell the Soldier that he/she will not perform first aid but will tell you what first aid action (give mouth-to-mouth resuscitation, bandage the wound, and so forth) he/she would take. After he/she has completed the checks ask him/her what else should be done. To test step 8, ask the Soldier what should he be doing while evacuating an unconscious casualty.

References:
Required:
Related:

081-COM-1003

Perform First Aid to Clear an Object Stuck in the Throat of a Conscious Casualty

Conditions: You see a conscious casualty who is having a hard time breathing because something is stuck in his/her throat.

Standards: Clear the object from the casualty's throat. Give abdominal or chest thrusts until the casualty can talk and breathe normally, you are relieved by a qualified person, or the casualty becomes unconscious requiring mouth-to-mouth resuscitation.

Special Condition: None

Special Standards: None

Special Equipment: None

Cue: None

Note: N/A

Performance Steps

   1. Determine if the casualty needs help.

      a. If the casualty has a mild airway obstruction (able to speak or cough forcefully, may be wheezing between coughs), do not interfere except to encourage the casualty.

      b. If the casualty has a severe airway obstruction (poor air exchange and increased breathing difficulty, a silent cough, cyanosis, or inability to speak or breathe), continue with step 2.

Note: You can ask the casualty one question, "Are you choking?" If the casualty nods yes, help is needed.

CAUTION: Do not slap a choking casualty on the back. This may cause the object to go down the airway instead of out.

   2. Perform abdominal or chest thrusts.

Note: Abdominal thrusts should be used unless the victim is in the advanced stages of pregnancy, is very obese, or has a significant abdominal wound.

Note: Clearing a conscious casualty's airway obstruction can be performed with the casualty either standing or sitting.

      a. Abdominal thrusts.

         (1) Stand behind the casualty.

         (2) Wrap your arms around the casualty's waist.

(3) Make a fist with one hand.

(4) Place the thumb side of the fist against the abdomen slightly above the navel and well below the tip of the breastbone.

(5) Grasp the fist with the other hand.

(6) Give quick backward and upward thrusts.

Note: Each thrust should be a separate, distinct movement. Thrusts should be continued until the obstruction is expelled or the casualty becomes unconscious.

b. Chest thrusts.

(1) Stand behind the casualty.

(2) Wrap your arms under the casualty's armpits and around the chest.

(3) Make a fist with one hand.

(4) Place the thumb side of the fist on the middle of the breastbone.

(5) Grasp the fist with the other hand.

(6) Give backward thrusts.

Note: Each thrust should be performed slowly and distinctly with the intent of relieving the obstruction.

3. Continue to give abdominal or chest thrusts, as required. Give abdominal or chest thrusts until the obstruction is clear, you are relieved by a qualified person, or the casualty becomes unconscious.

Note: If the casualty becomes unconscious, lay him/her down and then start mouth-to-mouth resuscitation procedures.

4. If the obstruction is cleared, watch the casualty closely and check for other injuries, if necessary.

Evaluation Preparation:

Setup: For training and evaluation, use another Soldier to simulate a patient in shock.

Brief Soldier: Tell the Soldier the simulated patient requires first aid for shock to be given.

Performance Measures	GO	NO GO
1 Determined if the casualty needed help.	_____	_____
2 Performed abdominal or chest thrusts, as required.	_____	_____
3 Continued abdominal or chest thrusts, as required.	_____	_____

Performance Measures	GO	NO GO
4 If the obstruction was cleared, watched the casualty closely and checked for other injuries, if necessary.	___	___

Evaluation Guidance: Score the Soldier GO if all steps are passed. Score the Soldier NO-GO if any step is failed. If the Soldier fails any step,show what was done wrong and how to do it correctly.

References
Required: FM 4-25.11
Related:

081-COM-1005

Perform First Aid to Prevent or Control Shock

Conditions: You have a casualty that is displaying one or more symptoms of shock. You have a field jacket or a poncho. The casualty is breathing and there is no uncontrolled bleeding. Some iterations of this task should be performed in MOPP.

Standards: Apply measures to prevent or treat shock without causing further injury to the casualty.

Special Condition: None

Special Standards: None

Special Equipment: None

Safety Level: Low

MOPP: Sometimes

Cue: None

Note:

Performance Steps

1. Check the casualty for signs and symptoms of shock.
   a. Sweaty but cool skin.
   b. Pale skin.

    c. Restlessness or nervousness.

    d. Thirst.

    e. Severe bleeding.

    f. Confusion.

    g. Rapid breathing.

    h. Blotchy blue skin.

    i. Nausea and/or vomiting.

  2. Position the casualty.

    a. Move the casualty under a permanent or improvised shelter to shade him from direct sunlight.

    b. Lay the casualty on his back unless a sitting position will allow the casualty to breathe easier.

    c. Elevate the casualty's feet higher than the heart using a stable object so the feet will not fall.

WARNING

Do not loosen clothing if in a chemical area.

---

  3. Loosen clothing at the neck, waist, or anywhere it is binding.

  4. Prevent the casualty from getting chilled or overheated. Using a blanket or clothing, cover the casualty to avoid loss of body heat by wrapping completely around the casualty.

Note: Ensure no part of the casualty is touching the ground, as this increases loss of body heat.

  5. Calm and reassure the casualty.

    a. Take charge and show self-confidence.

    b. Assure the casualty that he/she is being taken care of.

  6. Watch the casualty closely for life-threatening conditions and check for other injuries, if necessary. Seek medical aid.

  7. Seek medical aid.

Evaluation Preparation:

Setup: For training and evaluation, use another Soldier to simulate a patient in shock.

Brief Soldier: Tell the Soldier to treat the casualty to prevent or control shock.

Performance Measures	GO	NO GO
1 Checked the casualty for signs and symptoms of shock.	____	____

Performance Measures	GO	NO GO
2 Positioned casualty correctly.	_____	_____
3 Loosened clothing at the neck, waist, or anywhere it was binding.	_____	_____
4 Prevented the casualty from chilling or overheating.	_____	_____
5 Calmed and reassured the casualty.	_____	_____
6 Watched the casualty closely for life-threatening conditions and checked for other injuries, if necessary. Sought medical aid.	_____	_____
7 Sought Medical Aid.	_____	_____

Evaluation Guidance: Score each Soldier according to the performance measures. Unless otherwise stated in the task summary, the Soldier must pass all performance measures to be scored GO. If the Soldier fails any steps, show the Soldier what was done wrong and how to do the task correctly.

Environment: Environmental protection is not just the law but the right thing to do. It is a continual process and starts with deliberate planning. Always be alert to ways to protect our environment during training and missions. In doing so, you will contribute to the sustainment of our training resources while protecting people and the environment from harmful effects. Refer to FM 3-34.5 Environmental Considerations and GTA 05-08-002 ENVIRONMENTAL-RELATED RISK ASSESSMENT. Environmental protection is not just the law but the right thing to do. It is a continual process and starts with deliberate planning. Always be alert to ways to protect our environment during training and missions. In doing so, you will contribute to the sustainment of our training resources while protecting people and the environment from harmful effects. Refer to FM 3-34.5 Environmental Considerations and GTA 05-08-002 ENVIRONMENTAL-RELATED RISK ASSESSMENT.

Safety: In a training environment, leaders must perform a risk assessment in accordance with ATP 5-19, Risk Management. Leaders will complete a DD Form 2977 DELIBERATE RISK ASSESSMENT WORKSHEET during the planning and completion of each task and sub-task by assessing mission, enemy, terrain and weather, troops and support available-time available and civil considerations, (METT-TC). Note: During MOPP training, leaders must ensure personnel are monitored for potential heat injury. Local policies and procedures must be followed during times of increased heat category in order to avoid heat related injury. Consider the MOPP work/rest cycles and water replacement guidelines IAW FM 3-11.4, Multiservice Tactics, Techniques, and Procedures for Nuclear, Biological, and Chemical (NBC) Protection, FM 3-11.5, Multiservice Tactics, Techniques, and Procedures for Chemical, Biological, Radiological, and Nuclear Decontamination.

References:
Required:
Related: FM 4-25.11

081-COM-1023

Open An Airway

Conditions: You see an adult casualty who is unconscious and does not appear to be breathing. You are not in a combat situation or chemical environment. You will need a nasopharyngeal airway (NPA).

This task should not be trained in MOPP.

Standards: Take appropriate action, in the correct sequence, to open the airway.

Special Condition: None

Special Standards: None

Safety Level: Low

Special Equipment:

MOPP: Never

Note:

Performance Steps
WARNING The casualty should be carefully rolled as a whole, so the body does not twist.

1. Roll the casualty onto his/her back, if necessary, and place him/her on a hard, flat surface.

   a. Kneel beside the casualty.

   b. Raise the near arm and straighten it out above the head.

   c. Adjust the legs so they are together and straight or nearly straight.

   d. Place one hand on the back of the casualty's head and neck.

   e. Grasp the casualty under the arm with the free hand.

   f. Pull steadily and evenly toward yourself, keeping the head and neck in line with the torso.

   g. Roll the casualty as a single unit.

   h. Place the casualty's arms at his/her sides.

Cue: Casualty is unconscious, does not appear to be breathing, and is lying on his or her back.

2. Open the airway.

Note: If foreign material or vomit is in the mouth, remove it as quickly as possible.

> **CAUTION**
>
> Do NOT use this method if a spinal or neck injury is suspected.

   a. Head-tilt/chin-lift method.

   (1) Kneel at the level of the casualty's shoulders.

   (2) Place one hand on the casualty's forehead and apply firm, backward pressure with the palm to tilt the head back.

   (3) Place the fingertips of the other hand under the bony part of the lower jaw and lift, bringing the chin forward.

Note: Do NOT use the thumb to lift.

Note: Do NOT completely close the casualty's mouth.

CAUTION: Do NOT press deeply into the soft tissue under the chin with the fingers.

> **CAUTION**
>
> Use this method if a spinal or neck injury is suspected.
>
> *Note*: If you are unable to maintain an airway after the second attempt, use the head-tilt/chin-lift method.

   b. Jaw-thrust method.

   (1) Kneel above the casualty's head (looking toward the casualty's feet).

   (2) Rest your elbows on the ground or floor.

   (3) Place one hand on each side of the casualty's lower jaw at the angle of the jaw, below the ears.

   (4) Stabilize the casualty's head with your forearms.

   (5) Use the index fingers to push the angles of the casualty's lower jaw forward.

Note: If the casualty's lips are still closed after the jaw has been moved forward, use your thumbs to retract the lower lip and allow air to enter the casualty's mouth.

CAUTION: Do not tilt or rotate the casualty's head.

   3. Check for breathing.

      a. While maintaining the open airway position, place an ear over the casualty's mouth and nose, looking toward the chest and stomach.

      b. Look for the chest to rise and fall.

      c. Listen for air escaping during exhalation.

      d. Feel for the flow of air on the side of your face.

      e. Count the number of respirations for 15 seconds.

      f. Take appropriate action.

CAUTION

Do NOT use the NPA if there is clear fluid (cerebrospinal fluid-CSF) coming from the ears or nose. This may indicate a skull fracture.

---

     (1) If the casualty is unconscious, if respiratory rate is less than 2 in 15 seconds, and/or if the casualty is making snoring or gurgling sounds, insert an NPA.

       (a) Keep the casualty in a face-up position.

       (b) Lubricate the tube of the NPA with water.

       (c) Push the tip of the casualty's nose upward gently.

       (d) Position the tube of the NPA so that the bevel (pointed end) of the NPA faces toward the septum (the partition inside the nose that separates the nostrils).

Note: Most NPAs are designed to be placed in the right nostril.

CAUTION

Never force the NPA into the casualty's nostril. If resistance is met, pull the tube out and attempt to insert it in the other nostril. If neither nostril will accommodate the NPA, place the casualty in the recovery position.

       (e) Insert the NPA into the nostril and advance it until the flange rests against the nostril.

       (f) Place the casualty in the recovery position by rolling him/her as a single unit onto his/her side, placing the hand of his/her upper arm under his/her chin, and flexing his/her upper leg.

       (g) Watch the casualty closely for life-threatening conditions and check for other injuries, if necessary. Seek medical aid.

     (2) If the casualty is not breathing seek medical aid.

Note: If the casualty resumes breathing at any time during this procedure, the airway should be kept open and the casualty should be monitored. If the casualty continues to breathe, he/she should be transported to medical aid in accordance with the tactical situation.

Evaluation Preparation:

Setup: For training and testing, you must use a resuscitation training mannequin (DVC 08-15). Have a bottle of alcohol and swabs or cotton available. Place the mannequin on the floor and alcohol and cotton balls on the table. Clean the mannequin's nose and mouth before each Soldier is evaluated. If a mannequin that is capable of testing insertion of an NPA is available, use it to test step 3b.

Brief Soldier: Tell the Soldier to do, in order, all necessary steps to restore breathing. For step 3b, tell the Soldier that the casualty's breathing rate is slow, and have him show you (on a mannequin) or tell you what he would do to insert an NPA.

Performance Measures	GO	NO GO
1 Rolled the casualty onto his/her back, if necessary, and placed him/her on a hard, flat surface.	___	___
2 Opened the airway.	___	___
3 Checked for breathing.	___	___

Environment: Environmental protection is not just the law but the right thing to do. It is a continual process and starts with deliberate planning. Always be alert to ways to protect our environment during training and missions. In doing so, you will contribute to the sustainment of our training resources while protecting people and the environment from harmful effects. Refer to FM 3-34.5 Environmental Considerations and GTA 05-08-002 ENVIRONMENTAL-RELATED RISK ASSESSMENT.

Safety: In a training environment, leaders must perform a risk assessment in accordance with ATP 5-19, Risk Management. Leaders will complete a DD Form 2977 DELIBERATE RISK ASSESSMENT WORKSHEET during the planning and completion of each task and sub-task by assessing mission, enemy, terrain and weather, troops and support available-time available and civil considerations, (METT-TC). Note: During MOPP training, leaders must ensure personnel are monitored for potential heat injury. Local policies and procedures must be followed during times of increased heat category in order to avoid heat related injury. Consider the MOPP work/rest cycles and water replacement guidelines IAW FM 3-11.4, Multiservice Tactics, Techniques, and Procedures for Nuclear, Biological, and Chemical (NBC) Protection, FM 3-11.5, Multiservice Tactics, Techniques, and Procedures for Chemical, Biological, Radiological, and Nuclear Decontamination.

References:
Required:
Related: FM 4-25.11

References:

081-COM-1032

Perform First Aid for Bleeding of an Extremity

Conditions: You have a casualty who has a bleeding wound of the arm or leg. The casualty is breathing. You will need the casualty's emergency bandage, Kerlix field dressing, materials to improvise a pressure dressing (wadding and cravat or strip of cloth), and combat application tourniquet (C-A-T).

Some iterations of this task should be performed in MOPP.

Standards: Control bleeding from the wound without causing further harm to the casualty.

Special Condition: None

Special Standards: None

Special Equipment:

MOPP: Sometimes

Cue: None

Note: None

Performance Steps

> **CAUTION**
> All body fluids should be considered potentially infectious. Always observe body substance isolation (BSI) precautions by wearing gloves and eye protection as a minimal standard of protection. In severe cases, you should wear gloves, eye protection, gown and shoe covers to protect yourself of splashes, projectile fluids, spurting fluids or splashes onto your clothing and foot wear.

1. Determine if the bleeding is life threatening. If bleeding is life threatening, immediately apply a CAT tourniquet. See step #4.
Note: If in a tactical environment, evaluate a casualty (See task 081-COM-1001)

The three methods of controlling external bleeding are direct pressure, pressure dressing, and tourniquet.

CAUTION

Once bleeding has been controlled, it is important to check a distal pulse to make sure that the dressing has not been applied too tightly. If a pulse is not palpable, adjust the dressing to re-establish circulation.

---

2. If bleeding is not life threatening, apply direct pressure.

a. Expose the wound.

b. Place a sterile gauze or dressing over the injury site and apply fingertips, palm or entire surface of one hand and apply direct pressure.

c. Pack large, gaping wounds with sterile gauze and apply direct pressure.

WARNING

The emergency bandage must be loosened if the skin distal to the injury becomes cool, blue, numb, or pulseless.

CAUTION

---

3. Apply the pressure dressing (casualty's emergency bandage).

a. Open the plastic dressing package.

b. Apply the dressing, white (sterile, non-adherent pad) side down, directly over the wound.

c. Wrap the elastic tail (bandage) around the extremity and run the tail through the plastic pressure bar.

d. Reverse the tail while applying pressure and continue to wrap the remainder of the tail around the extremity, continuing to apply pressure directly over the wound.

e. Secure the plastic closure bar to the last turn of the wrap.

f. Check the emergency bandage to make sure that it is applied firmly enough to prevent slipping without causing a tourniquet-like effect.

CAUTION

In combat, while under enemy fire, a tourniquet is the primary means to control bleeding. It allows the individual, his battle buddy, or the combat medic to quickly control life threatening hemorrhage until the casualty can be moved away from the firefight. Always treat life threatening hemorrhage while you and the casualty are behind cover.

---

4. Apply a Combat Application Tourniquet (C-A-T).

a. Pull the free end of the self-adhering band through the buckle and route through the friction adapter buckle.

b. Place combat application tourniquet (C-A-T), 2-3 inches above the wound on the injured extremity.

c. Pull the self-adhering band tight around the extremity and fasten it back on itself as tightly as possible.

d. Twist the windlass until the bleeding stops.

e. Lock the windlass in place within the windlass clip.

f. Secure the windlass with the windlass strap.

g. Assess for absence of a distal pulse.

h. Place a "T" and the time of the application on the casualty with a marker.

i. Secure the C-A-T in place with tape.

5. Initiate treatment for shock as needed. (See task 081-COM-1005).

6. Record treatment given on the DD Form 1380, US Field Medical Card (FMC) or DA Form 7656, Tactical Combat Casualty Care (TCCC) Card.

7. Seek medical aid.

Evaluation Preparation:

Setup: For training and evaluation, use another Soldier to simulate a casualty with a bleeding extremity.

Brief Soldier: Tell the Soldier to treat the casualty with extremity bleeding.

	Performance Measures	GO	NO GO
1	Determined if the bleeding was life threatening. If bleeding was life threatening, immediately applied a C-A-T tourniquet. See step #4.	___	___
2	If bleeding was not life threatening, applied direct pressure.	___	___
3	Applied the pressure dressing (casualty's emergency bandage).	___	___
4	Applied a Combat Application Tourniquet (C-A-T).	___	___
5	Initiated treatment for shock as needed. (See task 081-COM-1005).	___	___
6	Recorded treatment given on the DD Form 1380, Field Medical Card (FMC) or DA Form 7656, Tactical Combat Casualty Care (TCCC) Card.	___	___
7	Sought medical aid.	___	___

Evaluation Guidance: Score each Soldier according to the performance measures. Unless otherwise stated in the task summary, the Soldier must pass all performance measures to be scored GO. If the Soldier fails any steps, show the Soldier what was done wrong and how to do the task correctly.

Environment: Environmental protection is not just the law but the right thing to do. It is a continual process and starts with deliberate planning. Always be alert to ways to protect our environment during training and missions. In doing so, you will contribute to the sustainment of our training resources while protecting people and the environment from harmful effects. Refer to FM 3-34.5 Environmental Considerations and GTA 05-08-002 ENVIRONMENTAL-RELATED RISK ASSESSMENT. Environmental protection is not just the law but the right thing to do. It is a continual process and starts with deliberate planning. Always be alert to ways to protect our environment during training and missions. In doing so, you will contribute to the sustainment of our training resources while protecting people and the environment from harmful effects. Refer to FM 3-34.5 Environmental Considerations and GTA 05-08-002 ENVIRONMENTAL-RELATED RISK ASSESSMENT.

Safety: In a training environment, leaders must perform a risk assessment in accordance with ATP 5-19, Risk Management. Leaders will complete a DD Form 2977 DELIBERATE RISK ASSESSMENT WORKSHEET during the planning and completion of each task and sub-task by assessing mission, enemy, terrain and weather, troops and support available-time available and civil considerations, (METT-TC). Note: During MOPP training, leaders must ensure personnel are monitored for potential heat injury. Local policies and procedures must be followed during times of increased heat category in order to avoid heat related injury. Consider the MOPP work/rest cycles and water replacement guidelines IAW FM 3-11.4, Multiservice Tactics, Techniques, and Procedures for Nuclear, Biological, and Chemical (NBC) Protection, FM 3-11.5, Multiservice Tactics, Techniques, and Procedures for Chemical, Biological, Radiological, and Nuclear Decontamination.

References:
Required:
Related: DA FORM 7656, Tactical Combat Casualty Care, DD FORM 1380

081-COM-1046

Transport a Casualty

Conditions: You have a casualty who has received treatment and requires movement and/or evacuation from a vehicle and placement on a transportation platform. You may have assistance from other Soldiers. You will need materials to improvise a litter (poncho, shirts, or jackets, and poles or tree limbs), a SKED or Talon litter, and a vehicle or replicated platform to load patients onto.

Some iterations of this task should be performed in MOPP.

Standards: Transport the casualty using an appropriate carry or litter without dropping or causing further injury to the casualty.

Special Condition: None

Special Standards: None

Special Equipment: None

Safety Level: Low

MOPP: Sometimes

Cue: A casualty must be moved.

Note: N/A

Performance Steps

> **WARNING**
>
> If the casualty was involved in a vehicle crash you should always consider that he/she may have a spinal injury. Unless there is an immediate life-threatening situation (such as fire, explosion), do NOT move the casualty with a suspected back or neck injury. Seek medical personnel for guidance on how to transport the casualty.

1. Remove a casualty from a vehicle, if necessary.
    a. Laterally.
        (1) With the assistance of another Soldier grasp the casualty's arms and legs.
        (2) While stabilizing the casualty's head and neck as much as possible, lift the casualty free of the vehicle and move him/her to a safe place on the ground.
Note: If medical personnel are available, they may stabilize the casualty's head, neck, and upper body with a special board or splint.
    b. Upward.
Note: You may have to remove a casualty upward from a vehicle; for example, from the passenger compartment of a wheeled vehicle lying on its side or from the hatch of an armored vehicle sitting upright.
        (1) You may place a pistol belt or similar material around the casualty's chest to help pull him/her from the vehicle.
        (2) With the assistance of another Soldier inside the vehicle, draw the casualty upward using the pistol belt or similar material or by grasping his/her arms.

(3) While stabilizing the casualty's head and neck as much as possible, lift the casualty free of the vehicle and place him/her on the topmost side of the vehicle.

Note: If medical personnel are available, they may stabilize the casualty's head, neck, and upper body with a special board or splint.

(4) Depending on the situation, move the casualty from the topmost side of the vehicle to a safe place on the ground.

WARNING

Do NOT use manual carries to move a casualty with a neck or spine injury, unless a life-threatening hazard is in the immediate area. Seek medical personnel for guidance on how to move and transport the casualty.

---

2. Select an appropriate method to transport the casualty.

Note: The fireman's carry is the typical one-man carry practiced in training. However, in reality, with a fully equipped casualty, it is nearly impossible to lift a Soldier over your shoulder and move to cover quickly. It should be discouraged from being practiced and used.

a. Fireman's carry. Use for an unconscious or severely injured casualty.

CAUTION

Do NOT use the neck drag if the casualty has a broken arm or a suspected neck injury.

---

b. Neck drag. Use in combat, generally for short distances.

c. Cradle-drop drag. Use to move a casualty who cannot walk when being moved up or down stairs.

d. Use litters if materials are available, if the casualty must be moved a long distance, or if manual carries will cause further injury.

Cue: The appropriate type of carry has been selected.

3. Evacuate the casualty using a manual carry.

a. Fireman's carry.

(1) Kneel at the casualty's uninjured side.

(2) Place casualty's arms above his/her head.

(3) Cross the ankle on the injured side over the opposite ankle.

(4) Place one of your hands on the shoulder farther from you and your other hand on his/her hip or thigh.

(5) Roll the casualty toward you onto his/her abdomen.

(6) Straddle the casualty.

Note: This method is used if the rescuer believes that it is safer than the regular method due to the casualty's wounds. Care must be taken to keep the casualty's head from falling backward, resulting in a neck injury.

(7) Place your hands under the casualty's chest and lock them together.

(8) Lift the casualty to his/her knees as you move backward.

(9) Continue to move backward, thus straightening the casualty's legs and locking the knees.

(10) Walk forward, bringing the casualty to a standing position but tilted slightly backward to prevent the knees from buckling.

(11) Maintain constant support of the casualty with one arm. Free your other arm, quickly grasp his/her wrist, and raise the arm high.

(12) Instantly pass your head under the casualty's raised arm, releasing it as you pass under it.

(13) Move swiftly to face the casualty.

(14) Secure your arms around his/her waist.

(15) Immediately place your foot between his/her feet and spread them (approximately 6 to 8 inches apart).

(16) Again grasp the casualty's wrist and raise the arm high above your head.

(17) Bend down and pull the casualty's arm over and down your shoulder bringing his/her body across your shoulders. At the same time pass your arm between the legs.

(18) Grasp the casualty's wrist with one hand while placing your other hand on your knee for support.

(19) Rise with the casualty correctly positioned.

Note: Your other hand is free for use as needed.

WARNING

Do NOT use the neck drag if the casualty has a broken and/or fractured arm or a suspected neck injury. If the casualty is unconscious, protect his/her head from the ground.

---

    b. Neck drag.

(1) Place the casualty on his back, if not already there. [See steps 3a (1)-(5)]. (

(2) Tie the casualty's hands together at the wrists. (If conscious, the casualty may clasp his/her hands together around your neck.)

(3) Straddle the casualty in a kneeling face-to-face position.

(4) Loop the casualty's tied hands over and/or around your neck.

(5) Crawl forward, looking ahead, dragging the casualty with you.

    c. Cradle-drop drag.

(1) With the casualty lying on his/her back, kneel at the head.

(2) Slide your hands, palms up, under the casualty's shoulders.

(3) Get a firm hold under his/her armpits.

(4) Partially rise, supporting the casualty's head on one of your forearms.

Note: You may bring your elbows together and let the casualty's head rest on both of your forearms.

(5) With the casualty in a semisitting position, rise and drag the casualty backwards.

(6) Back down the steps (or up if appropriate), supporting the casualty's head and body and letting the hips and legs drop from step to step.

    4. Evacuate the casualty using a SKED litter.

     a. Prepare the SKED litter for transport.

(1) Remove the SKED from the pack and place on the ground.

(2) Unfasten the retainer strap.

(3) Step on the foot end of the SKED litter and unroll the SKED completely.

(4) Bend the SKED in half and back roll.

(5) Repeat with the opposite end of the litter so that the SKED litter lays flat.

(6) Point out the handholds, straps for the casualty, and dragline at the head of the litter.

b. Place and secure a casualty to a SKED litter.

(1) Place the SKED litter next to the casualty so that the head end of the litter is next to the casualty's head.

(2) Place the cross straps under the SKED litter.

(3) Log roll the casualty onto his side in a steady and even manner.

(4) Slide the SKED litter as far under the casualty as possible.

(5) Gently roll the casualty until he is again lying on his back with the litter beneath him.

(6) Slide the casualty to the middle of the SKED litter, keeping his spinal column as straight as possible.

(7) Pull out the straps from under the SKED litter.

(8) Bring the straps across the casualty.

(9) Lift the sides of the SKED litter and fasten the four cross straps to the buckles directly opposite the straps.

(10) Lift the foot portion of the SKED litter.

(11) Feed the foot straps over the casualty's lower extremities and through the unused grommets at the foot end of the SKED litter.

(2) Fastens the straps to the buckles.

(13) Check to make sure the casualty is secured to the SKED litter.

c. Lift the casualty.

Note: For a SKED litter, lift the sides of the SKED and fasten the four cross straps to the buckles directly opposite the straps. Lift the foot portion of the SKED and feed the foot straps through the unused grommets at the foot end of the SKED and fasten to the buckles.

(1) Using four Soldiers (two on each side), all facing the casualty's feet. Have each rescuer grab a handle with their inside hand.

(2) In one fluid motion on the command of "prepare to lift, lift" raise as a unit holding the casualty parallel and even.

5. Evacuate the casualty using a Talon litter.

a. Prepare a Talon litter for use.

(1) Remove the litter from the bag.

(2) Stand the litter upright and release buckles from the litter.

(3) Place the litter on the ground and completely extend it with the fabric side facing up.

(4) Keeping the litter as straight as possible, grab the handles and rotate them inward until all the hinges rotate and lock.

Note: This action is done best using two individuals on each end of the litter executing this step simultaneously.

(5) While maintaining the hinges in the locked position, apply firm, steady pressure on the spreader bar with your foot. Increase pressure with your foot until the spreader bar locks into place.

b. Place the casualty on the litter.

(1) Place the litter next to the casualty. Ensure that the head end of the litter is beside the head of the casualty.

(2) Log roll the casualty and slide the litter as far under him/her as possible. Gently roll the casualty down onto the litter.

(3) Slide the casualty to the center of the litter. Be sure to keep the spinal column as straight as possible.

c. Secure the casualty to the litter using litter straps or other available materials.

6. Evacuate the casualty using an improvised litter.

a. Use the poncho and two poles or limbs.

(1) Open the poncho and lay the two poles lengthwise across the center, forming three equal sections.

(2) Reach in, pull the hood up toward you, and lay it flat on the poncho.

(3) Fold one section of the poncho over the first pole.

(4) Fold the remaining section of the poncho over the second pole to the first pole.

b. Use shirts or jackets and two poles or limbs.

(1) Zipper closed two uniform jackets and turn them inside out, leaving the sleeves inside.

(2) Lay the jackets on the ground and pass the poles through the sleeves, leaving one at the top and one at the
bottom of the poles to support the casualty's whole body.

c. Place the casualty on the improvised litter.

(1) Lift the litter.

(2) Place the litter next to the casualty. Ensure the head end of the litter is adjacent to the head of the casualty.

(3) Slide the casualty to the center of the litter. Be sure to keep the spinal column as straight as possible.

(4) Secure the casualty to the litter using litter straps or other available materials.

7. Load casualties onto a military vehicle.

a. Ground ambulance.

Note: Ground ambulances have combat medics to take care of the casualties during evacuation. Follow any special instructions that they give for loading, securing, or unloading casualties.

(1) Make sure each litter casualty is secured to his litter. Use the litter straps when available.

(2) Load the most serious casualty last.

(3) Load the casualty head first (head in the direction of travel) rather than feet first.

(4) Make sure each litter is secured to the vehicle.

Note: Unload casualties in reverse order, most seriously injured casualty first.

b. Air ambulance.

Note: Air ambulances have combat medics to take care of the casualties during evacuation. Follow any special instructions that they give for loading, securing, or unloading casualties.

(1) Remain 50 yards from the helicopter until the litter squad is signaled to approach the aircraft.

WARNING

Never go around the rear of the UH-60 or UH-1 aircraft.

---

(2) Approach the aircraft in full view of the aircraft crew, maintaining visual confirmation that the crew is aware of the approach of the litter party. Ensure that the aircrew can continue to visually distinguish friendly from enemy personnel at all times. Maintain a low silhouette when approaching the aircraft.

(a) Approach UH-60/UH-1 aircraft from sides. Do not approach from the front or rear. If you must move to the opposite side of the aircraft, approach from the side to the skin of the aircraft. Then hug the skin of the aircraft, and move around the front of the aircraft to the other side.

(b) Approach CH-47/CH46 aircraft from the rear.

(c) Approach MH-53 aircraft from the sides to the rear ramp, avoiding the tail rotor.

(d) Approach nonstandard aircraft in full view of the crew, avoiding tail rotors, main rotors, and propellers.

(e) Approach high performance aircraft (M/C-130/-141B/-17/-5B) from the rear, under the guidance of the aircraft loadmaster or the ground control party.

(3) Load the most seriously injured casualty last.

(4) Load the casualty who will occupy the upper berth first, and then load the next litter casualty immediately under the first casualty.

Note: This is done to keep the casualty from accidentally falling on another casualty if his litter is dropped before it is secured.

(5) When casualties are placed lengthwise, position them with their heads toward the direction of travel.

(6) Make sure each litter casualty is secured to his litter

(7) Make sure each litter is secured to the aircraft.

Note: Unload casualties in reverse order, most seriously injured casualty first.

c. Ground military vehicles.

Note: Nonmedical military vehicles may be used to evacuate casualties when no medical evacuation vehicles area available.

Note: If medical personnel are present, follow their instructions for loading, securing, and unloading casualties.

(1) When loading casualties into the vehicle, load the most seriously injured casualty last.

(2) When a casualty is placed lengthwise, load the casualty with his head pointing forward, toward the direction of travel.

(3) Ensure each litter casualty is secured to the litter. Use litter straps, if available.

(4) Secure each litter to the vehicle as it is loaded into place. Make sure each litter is secured.

Note: Unload casualties in reverse order, most seriously injured casualty first

---

Evaluation Preparation:

Setup: For training and evaluation, use other Soldiers to be simulated casualties to be transported. Place Soldiers in both vehicles and on the ground for transport. Have at least one tactical vehicle available for loading, or at least a large platform area that can accommodate several litter casualties.

Brief Soldier: Tell the Soldier the simulated casualties require movement to the evacuation platform.

Performance Measures	GO	NO GO
1 Removed the casualty from a vehicle, if necessary.	_____	_____
2 Selected an appropriate method of transporting the casualty.	_____	_____
3 Evacuated the casualty using a manual carry.	_____	_____
4 Evacuated the casualty using a SKED litter.	_____	_____
5 Evacuated a casualty using a Talon litter.	_____	_____
6 Evacuated a casualty using an improvised litter(s).	_____	_____
7 Loaded casualties onto a military vehicle.	_____	_____

Evaluation Guidance: Score each Soldier according to the performance measures. Unless otherwise stated in the task summary, the Soldier must pass all performance measures to be scored GO. If the Soldier fails any steps, show the Soldier what was done wrong and how to do the task correctly.

Environment: Environmental protection is not just the law but the right thing to do. It is a continual process and starts with deliberate planning. Always be alert to ways to protect our environment during training and missions. In doing so, you will contribute to the sustainment of our training resources while protecting people and the environment from harmful effects. Refer to FM 3-34.5 Environmental Considerations and GTA 05-08-002 ENVIRONMENTAL-RELATED RISK ASSESSMENT.

Safety: In a training environment, leaders must perform a risk assessment in accordance with ATP 5-19, Risk Management. Leaders will complete a DD Form 2977 DELIBERATE RISK ASSESSMENT WORKSHEET during the planning and completion of each task and sub-task by assessing mission, enemy, terrain and weather, troops and support available-time available and civil considerations, (METT-TC). Note: During MOPP training, leaders must ensure personnel are monitored for potential heat injury. Local policies and procedures must be followed during times of increased heat category in order to avoid heat related injury. Consider the MOPP work/rest cycles and water replacement guidelines IAW FM 3-11.4, Multiservice Tactics, Techniques, and Procedures for Nuclear, Biological, and Chemical (NBC) Protection, FM 3-11.5, Multiservice Tactics, Techniques, and Procedures for Chemical, Biological, Radiological, and Nuclear Decontamination.

References:
Required:
Related: FM 4-25.11, ATP 4-25.13, ATP 4-02.2

081-COM-1007

Perform First Aid for Burns

Conditions: You have a casualty who has a burn injury. You will need the casualty's emergency bandage or field dressing and canteen of water. Some iterations of this task should be performed in MOPP.

Standards: Give first aid for a burn without causing further injury to the casualty.

Special Condition: None

Special Standards: None

Special Equipment: None

Safety Level: Low

MOPP: Sometimes

Cue: None

Note: None

Performance Steps

1. Eliminate the source of the burn.

> **CAUTION**
>
> Synthetic materials, such as nylon, may melt and cause further injury.

    a. Thermal burns. Remove the casualty from the source of the burn. If the casualty's clothing is on fire, cover the casualty with a field jacket or any large piece of nonsynthetic material and roll him/her on the ground to put out the flames.

> **WARNING**
>
> Do not touch the casualty or the electrical source with your bare hands. You will be injured too!
>
> **WARNING:** High voltage electrical burns from an electrical source or lightning may cause temporary unconsciousness, difficulties in breathing, or difficulties with the heart (irregular heartbeat).

    b. Electrical burns. If the casualty is in contact with an electrical source, turn the electricity off, if the switch is nearby. If the electricity cannot be turned off, use any nonconductive material (rope, clothing, or dry wood) to drag the casualty away from the source.

> **WARNING**
>
> Blisters caused by a blister agent are actually burns. Do not try to decontaminate skin where blisters have already formed. If blisters have not formed, decontaminate the skin.

    c. Chemical burns.

        (1) Remove liquid chemicals from the burned casualty by flushing with as much water as possible.

        (2) Remove dry chemicals by carefully brushing them off with a clean, dry cloth. If large amounts of water are available, flush the area. Otherwise, do not apply water.

        (3) Smother burning white phosphorus with water, a wet cloth, or wet mud. Keep the area covered with the wet material.

    d. Laser burns. Move the casualty away from the source while avoiding eye contact with the beam source. If possible, wear appropriate laser eye protection.

Note: After the casualty has been removed from the source of the burn, continually monitor the casualty for conditions that may require basic lifesaving measures.

> **WARNING**
>
> Do NOT uncover the wound in a chemical environment. Exposure could cause additional harm.

2. Uncover the burn.

> **WARNING**
>
> Do NOT attempt to remove clothing that is stuck to the wound. Additional harm could result.

    a. Cut clothing covering the burned area.

> **CAUTION**
>
> Do not pull clothing over the burns.

    b. Gently lift away clothing covering the burned area.

    c. If the casualty's hand(s) or wrist(s) have been burned, remove jewelry (rings, watches) and place them in his/her pockets.

3. Apply the casualty's dry, sterile dressing directly over the wound.

Note: If the burn is caused by white phosphorus, the dressing must be wet.

CAUTION:

Do not place the dressing over the face or genital area.

Do not break the blisters.

Do not apply grease or ointments to the burns.

    a. Apply the dressing/pad, white side down, directly over the wound.

    b. Wrap the tails (or the elastic bandage) so that the dressing/pad is covered.

    c. For a field dressing, tie the tails into a nonslip knot over the outer edge of the dressing, not over the wound. For an emergency bandage, secure the hooking ends of the closure bar into the elastic bandage.

    d. Check to ensure that the dressing is applied lightly over the burn but firmly enough to prevent slipping.

Note: If the casualty is conscious and not nauseated, give him/her small amounts of water to drink.

4. Watch the casualty closely for life-threatening conditions, check for other injuries (if necessary), and treat for shock. Seek medical aid.

5. Seek medical aid.

Evaluation Preparation:

Setup: For training and evaluation, use another Soldier to simulate a casualty with a burn injury.

Brief Soldier: Tell the Soldier to treat the casualty with a burn injury.

Performance Measures	GO	NO GO
1 Eliminated the source of the burn.	_____	_____
2 Uncovered the burn, unless clothing was stuck to the wound or in a chemical environment.	_____	_____
3 Applied the dressing/pad directly over the wound.	_____	_____
4 Covered the edges of the dressing/pad.	_____	_____
5 Properly secured the bandage.	_____	_____
6 Applied the dressing lightly over the burn but firmly enough to prevent slipping.	_____	_____
7 Watched the casualty closely for life-threatening conditions, checked for other injuries (if necessary), and treated for shock.	_____	_____
8 Sought medical aid.	_____	_____

Evaluation Guidance: Score each Soldier according to the performance measures. Unless otherwise stated in the task summary, the Soldier must pass all performance measures to be scored GO. If the Soldier fails any steps, show the Soldier what was done wrong and how to do the task correctly.

Environment: Environmental protection is not just the law but the right thing to do. It is a continual process and starts with deliberate planning. Always be alert to ways to protect our environment during training and missions. In doing so, you will contribute to the sustainment of our training resources while protecting people and the environment from harmful effects. Refer to FM 3-34.5 Environmental Considerations and GTA 05-08-002 ENVIRONMENTAL-RELATED RISK ASSESSMENT. Environmental protection is not just the law but the right thing to do. It is a continual process and starts with deliberate planning. Always be alert to ways to protect our environment during training and missions. In doing so, you will contribute to the sustainment of our training resources while protecting people and the environment from harmful effects. Refer to FM 3-34.5 Environmental Considerations and GTA 05-08-002 ENVIRONMENTAL-RELATED RISK ASSESSMENT.

Safety: In a training environment, leaders must perform a risk assessment in accordance with ATP 5-19, Risk Management. Leaders will complete a DD Form 2977 DELIBERATE RISK ASSESSMENT WORKSHEET during the planning and completion of each task and sub-task by assessing mission, enemy, terrain and weather, troops and support available-time available and civil considerations, (METT-TC). Note: During MOPP training, leaders must ensure personnel are monitored for potential heat injury. Local policies and procedures must be followed during times of increased heat category in order to avoid heat related injury. Consider the MOPP work/rest cycles and water replacement guidelines IAW FM 3-11.4, Multiservice Tactics, Techniques, and Procedures for Nuclear, Biological, and Chemical (NBC) Protection, FM 3-11.5, Multiservice Tactics, Techniques, and Procedures for Chemical, Biological, Radiological, and Nuclear Decontamination.

References:
Required:
Related: FM 4-25.11

081-COM-1026

Perform First Aid for an Open Chest Wound

Condition: You have a casualty who has an open chest injury. The casualty is breathing and has no life threatening bleeding. You will need a commercial occlusive chest seal or a wrapper from a dressing, tape and dressing material.

Some iterations of this task should be performed in MOPP.

Standard: Perform first aid to the open chest wound without causing further injury to the casualty.

Special Condition: None

Safety Level: Low

MOPP: Sometimes

Performance Steps

> **CAUTION:**
> Removing stuck clothing or uncovering the wound in a chemical environment could cause additional harm.

Cue: You have a casualty with an open chest wound who requires first aid.

1. Uncover the wound (unless clothing is stuck to the wound or you are in a chemical environment).

Note: If you are not sure if the wound has penetrated the chest wall completely, treat the wound as though it were an open chest wound.

If multiple wounds are found at once, treat the largest one first.

Cue: The wound was uncovered.

2. Place gloved hand or back of hand over open chest wound to create temporary seal.

Note: Since air can pass through most dressings and bandages, you must seal the open chest wound with a commercial chest seal, plastic, cellophane, or other nonporous, airtight material to prevent air from entering the chest.

3. Apply airtight material over the wound.

a. Fully open the outer wrapper of the casualty's dressing, commercial chest seal or other airtight material.

b. Place the inner surface of the outer wrapper or other airtight material directly over the wound after the casualty exhales completely. Edges of the airtight material must extend 2 inches beyond the edges of the wound.

Note: When applying the airtight material, do not touch the inner surface.

c. Apply two inch tape (found in IFAK) to all four sides of the material securing it to the casualty's chest.

4. Check for exit wound or other open chest injuries.

Note: If exit wound or other open chest injuries are found, perform same steps as for entrance wound.

5. Apply the casualty's emergency dressing over air tight material.

a. Apply the dressing/pad, white side down, directly over the airtight material.

b. Have the casualty breathe normally.

c. Maintain pressure on the dressing while you wrap the tails (or elastic bandage) around the body and back to the starting point.

d. Pass the tail through the plastic pressure device, reverse the tail while applying pressure, continue to wrap the tail around the body, and secure the plastic fastening clip to the last turn of the wrap.

e. Ensure that the dressing is secured without interfering with breathing.

6. Position the casualty.

a. Place a conscious casualty in the sitting position or on his side (recovery position) with his injured side next to the ground.

Note: If the casualty is having difficulty breathing, place him in a position of comfort to ease breathing.

b. Place an unconscious casualty in the recovery position on the injured side.

7. Monitor the casualty closely for life-threatening conditions, check for other injuries (if necessary), and treat for shock.

8. Seek medical aid.

Evaluation Guidance: Tell the Soldier to do, in order, all necessary first aid steps to treat the casualty's wound. When testing step 1, you can vary the test by telling the Soldier that clothing is stuck to the wound or that a chemical environment exists.

Evaluation Preparation: Have a Soldier act as the casualty. Use a moulage kit or otherwise simulate the chest wound.

Performance Measures	GO	NO GO
1 Uncovered the wound.	_____	_____
2 Placed gloved hand or back of hand over chest wound to create temporary seal.	_____	_____
3 Applied airtight material to seal the wound.	_____	_____
4 Checked for exit wound or other open chest injuries.	_____	_____
5 Applied casualty's emergency dressing over the airtight material.	_____	_____
6 Positioned the casualty.	_____	_____
7 Monitored the casualty closely for life-threatening conditions, checked for other injuries (if necessary), and treated for shock.	_____	_____
8 Sought medical aid.	_____	_____

Environment: Environmental protection is not just the law but the right thing to do. It is a continual process and starts with deliberate planning. Always be alert to ways to protect our environment during training and missions. In doing so, you will contribute to the sustainment of our training resources while protecting people and the environment from harmful effects. Refer to FM 3-34.5 Environmental Considerations and GTA 05-08-002 ENVIRONMENTAL-RELATED RISK ASSESSMENT. Environmental protection is not just the law but the right thing to do. It is a continual process and starts with deliberate planning. Always be alert to ways to protect our environment during training and missions. In doing so, you will contribute to the sustainment of our training resources while protecting people and the environment from harmful effects. Refer to FM 3-34.5 Environmental Considerations and GTA 05-08-002 ENVIRONMENTAL-RELATED RISK

ASSESSMENT.

Safety: In a training environment, leaders must perform a risk assessment in accordance with ATP 5-19, Risk Management. Leaders will complete a DD Form 2977 DELIBERATE RISK ASSESSMENT WORKSHEET during the planning and completion of each task and sub-task by assessing mission, enemy, terrain and weather, troops and support available-time available and civil considerations, (METT-TC). Note: During MOPP training, leaders must ensure personnel are monitored for potential heat injury. Local policies and procedures must be followed during times of increased heat category in order to avoid heat related injury. Consider the MOPP work/rest cycles and water replacement guidelines IAW FM 3-11.4, Multiservice Tactics, Techniques, and Procedures for Nuclear, Biological, and Chemical (NBC) Protection, FM 3-11.5, Multiservice Tactics, Techniques, and Procedures for Chemical, Biological, Radiological, and Nuclear Decontamination.

References:
Required: FM 4-25.11
Related:

081-COM-0101

Request Medical Evacuation

Conditions: You have a casualty requiring medical evacuation (MEDEVAC). You will need operational communications equipment, MEDEVAC request format, and unit signal operation instructions (SOI). Some iterations of this task should be performed in MOPP.

Some iterations of this task should be performed in MOPP.

Standards: Transmit a 9-Line MEDEVAC request, providing all necessary information as quickly as possible. Transmit, as a minimum, line numbers 1 through 5 during the initial contact with the evacuation unit. Transmit lines 6 through 9 while the aircraft or vehicle is en route, if not included during the initial contact. IAW ATP 4-02.2, Medical Evacuation.

Special Condition: None

Special Standards: None

Special Equipment:

Safety Level: Low

MOPP: Sometimes

Cue: None

Note: None

References:
Required:
Related:

Performance Steps

1. Collect all applicable information needed for the MEDEVAC request.

   a. Determine the grid coordinates for the pickup site. (See STP 21-1-SMCT, task 071-COM-1002.)

   b. Obtain radio frequency, call sign, and suffix.

   c. Obtain the number of patients and precedence.

   d. Determine the type of special equipment required.

   e. Determine the number and type (litter or ambulatory) of patients.

   f. Determine the security of the pickup site.

   g. Determine how the pickup site will be marked.

   h. Determine patient nationality and status.

   i. Obtain pickup site chemical, biological, radiological, and nuclear (CBRN) contamination information normally obtained from the senior person or medic.

Note: CBRN line 9 information is only included when contamination exists.

2. Record the gathered MEDEVAC information using the authorized brevity codes. (See table 081-COM-0101-1 and 081-COM-0101-2.)

Note: Unless the MEDEVAC information is transmitted over secure communication systems, it must be encrypted, except as noted in step 3b(1).

Table 081-COM-0101-1

LINE	ITEM	EXPLANATION	WHERE/HOW OBTAINED	WHO NORMALLY PROVIDES	REASON
1	Location of pickup site	Encrypt the grid coordinates of the pickup site. When using the DRYAD Numeral Cipher, the same "SET" line will be used to encrypt the grid zone letters and the coordinates. To preclude misunderstanding, a statement is made that grid zone letters are included in the message (unless unit SOP specifies its use at all times).	From map	Unit leader(s)	Required so evacuation vehicle knows where to pick up patient. Also, so that the unit coordinating the evacuation mission can plan the route for the evacuation vehicle (if the evacuation vehicle must pick up from more than one location).
2	Radio frequency, call sign, and suffix	Encrypt the frequency of the radio at the pickup site, not a relay frequency. The call sign (and suffix if used) of person to be contacted at the pickup site may be transmitted in the clear.	From SOI	RTO	Required so that evacuation vehicle can contact requesting unit while en route (obtain additional information or change in situation or directions).
3	Number of patients by precedence	Report only applicable information and encrypt the brevity codes. A - URGENT B - URGENT-SURG C - PRIORITY D - ROUTINE E - CONVENIENCE If two or more categories must be reported in the same request, insert the word "BREAK" between each category.	From evaluation of patient(s)	Medic or senior person present	Required by unit controlling vehicles to assist in prioritizing missions.
4	Special equipment required	Encrypt the applicable brevity codes. A - None B - Hoist C - Extraction equipment D - Ventilator	From evaluation of patient/situation	Medic or senior person present	Required so that the equipment can be placed on board the evacuation vehicle prior to the start of the mission.
5	Number of patients by type	Report only applicable information and encrypt the brevity code. If requesting medical evacuation for both types, insert the word "BREAK" between the litter entry and ambulatory entry. L + # of patients - Litter A + # of patients - Ambulatory (sitting)	From evaluation of patient(s)	Medic or senior person present	Required so that the appropriate number of evacuation vehicles may be dispatched to the pickup site. They should be configured to carry the patients requiring evacuation.
6	Security of pickup site (wartime)	N - No enemy troops in area P - Possibly enemy troops in area (approach with caution) E - Enemy troops in area (approach with caution) X - Enemy troops in area (armed escort required)	From evaluation of situation	Unit leader	Required to assist the evacuation crew in assessing the situation and determining if assistance is required. More definitive guidance can be furnished the evacuation vehicle while it is en route (specific location of enemy to assist an aircraft in planning its approach).

Table 081-COM-0101-2

LINE	ITEM	EXPLANATION	WHERE/HOW OBTAINED	WHO NORMALLY PROVIDES	REASON
6	Number and type of wound, injury, or illness (peacetime)	Specific information regarding patient wounds by type (gunshot or shrapnel). Report serious bleeding, along with patient's blood type, if known.	From evaluation of patient(s)	Medic or senior person present	Required to assist evacuation personnel in determining treatment and special equipment needed.
7	Method of marking pickup site	Encrypt the brevity codes. A - Panels B - Pyrotechnic signal C - Smoke signal D - None E - Other	Based on situation and availability of materials	Medic or senior person present	Required to assist the evacuation crew in identifying the specific location of the pickup. Note that the color of the panels or smoke should not be transmitted until the evacuation vehicle contacts the unit (just prior to its arrival). For security, the crew should identify the color and the unit verifies it.
8	Patient nationality and status	The number of patients in each category need not be transmitted. Encrypt only the applicable brevity codes. A - US military B - US citizen C - Non-US military D - Non-US citizen E - Enemy prisoner of war (EPW)	From evaluation of patient(s)	Medic or senior person present	Required to assist in planning for destination facilities and need for guards. Unit requesting support should ensure that there is an English-speaking representative at the pickup site.
9	CBRN contamination (wartime)	Include this line only when applicable. Encrypt the applicable brevity codes. C - Chemical B - Biological R - Radiological N - Nuclear	From situation	Medic or senior person present	Required to assist in planning for the mission (determine which evacuation vehicle will accomplish the mission and when it will be accomplished).
9	Terrain description (peacetime)	Include details of terrain features in and around proposed landing site. If possible, describe relationship of site to prominent terrain feature (lake, mountain, tower).	From area survey	Personnel present	Required to allow evacuation personnel to assess route/avenue of approach into area. Of particular importance if hoist operation is required.

a. Location of the pickup site (line 1).

b. Radio frequency, call sign, and suffix (line 2).

c. Numbers of patients by precedence (line 3).

(1) Encrypt this information using the following brevity codes: A=Urgent. B= Urgent Surgical. C= Priority. D=Routine. E= Convenience.

(2) If 2 or more categories are reported in same request, insert the word "break" between each category.

d. Special equipment required (line 4). Encrypt this information using the following brevity codes: A= None. B= Hoist. C= Extraction Equipment. D= Ventilator.

e. Number of patients by type (line 5). Encrypt this information using the following brevity codes: L+#: Number of litter patients. A+#: Number of ambulatory patients (able to walk or can walk with assistance).
Note: If requesting MEDEVAC for both types, insert the word "break" between the litter entry and the ambulatory entry.

f. Security of the pickup site (line 6- wartime). Encrypt this information using the following brevity codes: N= No enemy troops in area. P= Possibly enemy troops in area, approach with caution. E= Enemy troops in area, approach with caution. X= Enemy troops in area, armed escort required.

g. Number and type of wound, injury or illness (line 6- peacetime)

h. Method of marking the pickup site (line 7). Encrypt this information using the following brevity codes: A= Panels. B= Pyrotechnic signal. C= Smoke signal. D= None. E= Other.

i. Patient nationality and status (line 8). Encrypt this information using the following brevity codes: A= US Military. B=US Civilian. C= Non-US Military. D= Non-US Civilian. E= Enemy prisoner (EPW).

j. CBRN contamination (line 9). Encrypt this information using the following brevity codes: N= Nuclear or radiological. B= Biological. C= Chemical.

k. Terrain Description (line 9 - peacetime)

3. Transmit the MEDEVAC request. (See STP 21-1-SMCT, task 113-COM-1022.)

Note: Transmission may vary depending on individual experience level and situation.

a. Contact the unit that controls the evacuation assets.

(1) Make proper contact with the intended receiver. Use effective call sign and frequency assignments from the SOI.

(2) Give the following in the clear "I HAVE A MEDEVAC REQUEST;" wait one to three seconds for a response. If no response, repeat the statement.

b. Transmit the MEDEVAC information in the proper sequence.

(1) State all line item numbers in clear text. The call sign and suffix (if needed) in line 2 may be transmitted in the clear.text.

Note: Line numbers 1 through 5 must always be transmitted during the initial contact with the evacuation unit. Lines 6 through 9 may be transmitted while the aircraft or vehicle is en route.

(2) Follow the procedure provided in the explanation column of the MEDEVAC request format to transmit other required information. (See tables 081-COM-0101-1 and 081-COM-0101-2.)

(3) Pronounce letters and numbers according to appropriate radiotelephone procedures.

(4) End the transmission by stating "OVER."

(5) Keep the radio on and listen for additional instructions or contact from the evacuation unit.

4. Keep the radio on and listen for additional instructions or contact from the evacuation unit.

Evaluation Preparation:

Setup: For evaluation of this task, create a scenario and provide the Soldier information for the request as the Soldier requests it. You or an assistant will act as the radio contact at the evacuation unit during "transmission" of the request. Give a copy of the MEDEVAC request format to the Soldier.

Brief Soldier: Tell the Soldier to prepare and transmit a MEDEVAC request. State that the communication net is secure.

Performance Measures	GO	NO GO
1 Collected all information needed for the MEDEVAC request line items 1 through 9.	_____	_____
Note: Wartime procedures for line items 6 and 9 will be used.		
2 Recorded the information using the authorized brevity codes.	_____	_____
3 Transmitted the MEDEVAC request as quickly as possible, following appropriate radiotelephone procedures.	_____	_____
4 Kept the radio on, listening for additional instruction or contact from the evacuation unit.	_____	_____

Evaluation Guidance: Score the Soldier GO if all steps are passed. Score the Soldier NO-GO if any step is failed. If the Soldier fails any step,show what was done wrong and how to do it correctly.

Environment: Environmental protection is not just the law but the right thing to do. It is a continual process and starts with deliberate planning. Always be alert to ways to protect our environment during training and missions. In doing so, you will contribute to the sustainment of our training resources while protecting people and the environment from harmful effects. Refer to FM 3-34.5 Environmental Considerations and GTA 05-08-002 ENVIRONMENTAL-RELATED RISK ASSESSMENT. Environmental protection is not just the law but the right thing to do. It is a continual process and starts with deliberate planning. Always be alert to ways to protect our environment during training and missions. In doing so, you will contribute to the sustainment of our training resources while protecting people and the environment from harmful effects. Refer to FM 3-34.5 Environmental Considerations and GTA 05-08-002 ENVIRONMENTAL-RELATED RISK ASSESSMENT.

Safety: In a training environment, leaders must perform a risk assessment in accordance with ATP 5-19, Risk Management. Leaders will complete a DD Form 2977 DELIBERATE RISK ASSESSMENT WORKSHEET during the planning and completion of each task and sub-task by assessing mission, enemy, terrain and weather, troops and support available-time available and civil considerations, (METT-TC). Note: During MOPP training, leaders must ensure personnel are monitored for potential heat injury. Local policies and procedures must be followed during times of increased heat category in order to avoid heat related injury. Consider the MOPP work/rest cycles and water replacement guidelines IAW FM 3-11.4, Multiservice Tactics, Techniques, and Procedures for Nuclear, Biological, and Chemical (NBC) Protection, FM 3-11.5, Multiservice Tactics, Techniques, and Procedures for Chemical, Biological, Radiological, and Nuclear Decontamination.

References:
Required: ATP 4-02.2, ATP 4-25.13, STP 21-1-SMCT
Related: FM 6-02.53

052-COM-1270

React to an Improvised Explosive Device (IED) Attack (Located at https://www.us.army.mil/suite/doc/23838478) (UNCLASSIFIED//FOR OFFICIAL USE ONLY) (U//FOUO)

Conditions: This task is identified as FOUO, refer to DTMS or CAR to view

---

052-COM-1271

Identify visual Indicators of an Implosive Device (IED) (Located at https://www.us.army.mil/suite/doc/23838510) (UNCLASSIFIED//FOR OFFICIAL USE ONLY) (U//FOUO)

Conditions: This task is identified as FOUO, refer to DTMS or CAR to view

---

071-COM-0804

Perform Surveillance without the Aid of Electronic Device

Conditions: You are a member of a squad or team in a defensive position and must conduct surveillance within your assigned sector during both daylight and limited visibility (night).

Standards: Identify potential activity indicators and conduct a visual search of your assigned sector. Submit SALUTE reports, as required.

Special Condition: None

Special Standards: None

Special Equipment:

Cue:None

*Note:*None

Performance Steps

1. Identify potential activity indicators in sector (Figure 071-COM-0804-1).

SIGHT Look for--	SOUND Listen for--	TOUCH Feel for--	SMELL Smell for--
• Enemy personnel, vehicles, and aircraft	• Running engines or track sounds	• Warm coals and other materials in a fire	• Vehicle exhaust
• Sudden or unusual movement	• Voices	• Fresh tracks	• Burning petroleum products
• New local inhabitants	• Metallic sounds	• Age of food or trash	• Food cooking
• Smoke or dust	• Gunfire, by weapon type		• Aged food in trash
• Unusual movement of farm or wild animals	• Unusual calm or silence		• Human waste
• Unusual activity--or lack of activity--by local inhabitants, especially at times or places that are normally inactive or active	• Dismounted movement		
	• Aircraft		

OTHER CONSIDERATIONS		
• Vehicle or personnel tracks	Armed Elements	Locations of factional forces, mine fields, and potential threats.
• Movement of local inhabitants along uncleared routes, areas, or paths	Homes and Buildings	Condition of roofs, doors, windows, lights, power lines, water, sanitation, roads, bridges, crops, and livestock.
• Signs that the enemy has occupied the area	Infrastructure	Functioning stores, service stations, and so on.
• Evidence of changing trends in threats	People	Numbers, gender, age, residence or DPRE status, apparent health, clothing, daily activities, and leadership.
• Recently cut foliage	Contrast	Has anything changed? For example, are there new locks on buildings? Are windows boarded up or previously boarded up windows now open, indicating a change in how a building is expected to be used? Have buildings been defaced with graffiti?
• Muzzle flashes, lights, fires, or reflections		
• Unusual amount (too much or too little) of trash		

Figure 071-COM-0804-1. Potential Indicators.

2. Perform observation techniques of the sector.
    a. Conduct day observation.
        (1) Use rapid scan technique. (Figure 071-COM-0804-2).
Note: The rapid scan technique is used to detect obvious signs of enemy activity. It is usually the first method you will use.

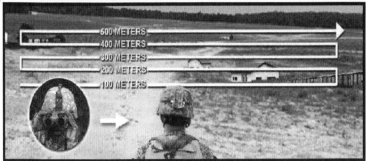

Figure 071-COM-0804-2. Rapid/Slow Scan.

      (a) Search a strip of terrain about 100 meters deep, from left-to-right, pausing at short intervals.

      (b) Search another 100-meter strip farther out, from right-to-left, overlapping the first strip scanned, pausing at short intervals.

      (c) Continue this method until the entire sector of fire has been searched.

      (2) Use slow scan technique.

Note: Slow scan search technique uses the same process as the rapid scan but much more deliberately; this means a slower, side-to-side movement and more frequent pauses.

      (3) Use detailed search technique paying attention to the following: (Figure 071-COM-0804-3).

Note: The detailed search, even more than the rapid or slow scan, depends on breaking a larger sector into smaller sectors to ensure everything is covered in detail and no possible enemy positions are overlooked.

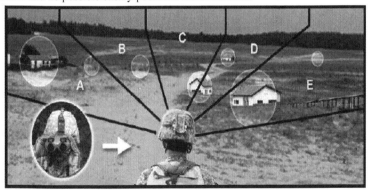

Figure 071-COM-0804-3. Detailed Search.

      (a) Likely enemy positions and suspected vehicle/dismounted avenues of approach.

(b) Target signatures, such as road junctions, hills, and lone buildings, located near prominent terrain features.

(c) Areas with cover and concealment, such as tree lines and draws.

b. Conduct limited visibility observation.

(1) Use dark adaptation technique.

(a) Stay in a dark area for about 30 minutes.

(b) Move into a red-light area for about 20 minutes followed by about 10 minutes in a dark area.

Note: The red-light method may save time by allowing you to get orders, check equipment, or do some other job before moving into darkness.

(2) Use night vision scan technique (Figure 071-COM-0804-4).

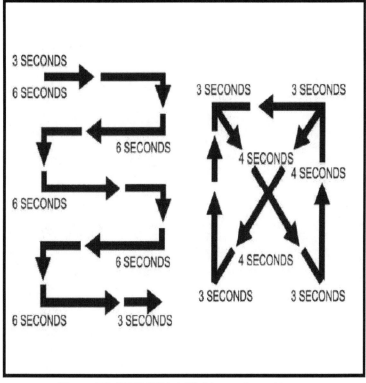

Figure 071-COM-0804-4. Night scanning patterns.

(a) Look from right to left or left to right using a slow, regular scanning movement.

(b) At night avoid looking directly at a faintly visible object when trying to confirm its presence.

(3) Use off center vision technique.

Note: The technique of viewing an object using central vision is ineffective at night due to the night blind spot that exist during low illumination. You must learn to use off-center vision.

  (a) View an object by looking 10 degrees above, below, or to either side of it rather than directly at it.

  (b) Shift your eyes from one off-center point to another.

  (c) Continue to pick-up the object in your peripheral field of vision.

3. Submit SALUTE report (Figure 071-COM-0804-5).

Line No.	Type Info	Description
1	(S)ize/Who	Expressed as a quantity and echelon or size. For example, report "10 enemy Infantrymen" (not "a rifle squad").
If multiple units are involved in the activity you are reporting, you can make multiple entries.		
2	(A)ctivity/What	Relate this line to the PIR being reported. Make it a concise bullet statement. Report what you saw the enemy doing, for example, "emplacing mines in the road."
3	(L)ocation/Where	This is generally a grid coordinate, and should include the 100,000-meter grid zone designator. The entry can also be an address, if appropriate, but still should include an eight-digit grid coordinate. If the reported activity involves movement, for example, advance or withdrawal, then the entry for location will include "from" and "to" entries. The route used goes under "Equipment/How."
4	(U)nit/Who	Identify who is performing the activity described in the "Activity/What" entry. Include the complete designation of a military unit, and give the name and other identifying information or features of civilians or insurgent groups.
5	(T)ime/When	For future events, give the DTG for when the activity will initiate. Report ongoing events as such. Report the time you saw the enemy activity, not the time you report it. Always report local or Zulu (Z) time.
6	(E)quipment/How	Clarify, complete, and expand on previous entries. Include information about equipment involved, tactics used, and any other essential elements of information (EEI) not already reported in the previous lines.

Figure 071-COM-0804-5. SALUTE Format.

Evaluation Preparation:

Setup: Provide the Soldier with the equipment and or materials described in the conditions statement.

Brief Soldier: Tell the Soldier what is expected of him by reviewing the task standards. Stress to the Soldier the importance of observing all cautions, warnings, and dangers to avoid injury to personnel and, if applicable, damage to equipment.

Performance Measures	GO	NO GO
1 Identified potential activity indicators in sector.	_____	_____
2 Performed observation techniques of the sector.	_____	_____
3 Submitted SALUTE report.	_____	_____

Evaluation Guidance: Score the Soldier GO if all performance measures are passed. Score the Soldier NO-GO if any performance measure is failed. If the Soldier scores a NO-GO, show the Soldier what was done wrong and how to do it correctly.

References:
Required:
Related: TC 3-21.75

301-COM-1050

Report Information of Potential Intelligence Value

WARNING
Do not wait until you have complete information to transmit. Even small amounts of information of critical tactical value may provide indicators of the threat's intentions.

Conditions: You are a Soldier with the responsibility to actively observe and provide concise accurate reports while in an area of operations. You are given information requirements, a means of communication (radio, wire, cable, or messenger) as prescribed in the unit's standard operating procedures (SOPs), required mission-specific equipment and a situation which requires you to immediately report information of critical tactical value.

Standards: Transmit information to the receiving authority in size, activity, location, unit, time and equipment (SALUTE) format to include significant terrain and weather conditions via the available means of communication. Information will be reported within 5 minutes after observation with six out of six SALUTE items correctly identified. Note: Your unit SOPs will specify the receiving authority. Examples of receiving authorities are company commander, team commander, company intelligence support team (CoIST), or S2 (Intelligence Officer [U.S. Army]) section.

Special Condition: None

Special Standards: None

Special Equipment:

Cue: None

*Note:*None

Performance Steps

1. Identify information concerning threat activity and significant terrain and weather conditions including-

a. Order of battle factors; for example, threat weapons systems, composition, and direction of movement.

Note: If you cannot identify a weapon system or vehicle by name, include a description of the equipment.

b. Military aspects of terrain; for example, observation and fields of fire, avenues of approach, key and decisive terrain, obstacles, and cover and concealment (OAKOC).

c. Weather factors; for example, severe weather, precipitation, trafficability, surface winds and gusts, and ground visibility.

Note: Use Spot Reports (Level 1 Report) to transmit information of immediate value. Transmit Spot Reports as rapidly and securely as possible. The SALUTE format is an aid for the observer to report the essential reporting elements. (You may precede each message segment of the Spot Report with the meaning of the acronym SALUTE.)

2. Draft message summary information in the SALUTE format.

a. S-Size. Report the number of personnel, vehicles, aircraft, or size of an object. Make an estimate if necessary.

b. A-Activity. Report detailed account of the detected element activity. Indicate the activity types or types and an amplifying sub-type if applicable.

(1) Attacking. (direction from)

(a) ADA. (engaging)

(b) Aircraft. (engaging) (rotary wing (RW), fixed wing (FW))

        (c) Ambush. (improvised explosive device (IED) (exploded), IED (unexploded), Sniper, Anti-armor, Other)

           (d) Indirect fire. (point of impact, point of origin)

           (e) Chemical, biological, radiological, and nuclear (CBRN)

      (2) Defending. (direction from)

      (3) Moving. (direction from)

      (4) Stationary.

      (5) Cache.

      (6) Civilian. (criminal acts, unrest, infrastructure damage)

      (7) Personnel recovery. (isolating event, observed signal)

      (8) Other. (Give name and description)

  c. L-Location. Report where you saw the activity. Include grid coordinates with Grid Zone Designator or reference from a known point including the distance and direction from the known point.

  d. U-Unit. Report the detected element unit, organization, or facility. Indicate the type of unit, organization, or facility detected. If it cannot be clearly identified, describe in as much detail as possible, including uniforms, vehicle markings, and other identifying information.

      (1) Conventional.

      (2) Irregular.

      (3) Coalition.

      (4) Host Nation.

      (5) Non-governmental Organization (NGO).

      (6) Civilian.

      (7) Facility.

  e. T-Time. Report the time and date the activity was observed, not the time you report it. Always report local or Zulu time.

  f. E-Equipment. Report all equipment associated with the activity, such as weapons, vehicles, tools. Add a narrative if necessary to clarify, describe, or explain the type of equipment. Provide nomenclature, type, and quantity of all equipment observed. If equipment cannot be clearly identified, describe in as much detail as possible.

      (1) Air Defense Artillery (ADA) (missile (MANPADS), missile (other), gun)

      (2) Artillery (gun (self propelled), gun (towed), missile or rocket, mortar)

      (3) Armored track vehicle (tank, APC, command and control (C2), engineer, transport, other)

      (4) Armored wheel vehicle (gun, APC, C2, engineer, transport, other)

      (5) Wheel vehicle (gun, C2, engineer, transport, other)

      (6) Infantry weapon (anti-armor missile, anti-armor gun, RPG, heavy machinegun, GL, small arms, other)

      (7) Aircraft (RW (attack helicopter (AH)), RW (utility helicopter (UH)), RW (observation helicopter), FW (attack), FW (transport), unmanned aircraft, other)

      (8) Mine or IED (buried, surface, VBIED, PBIED, other)

      (9) CBRN

(10) Supplies (Class III, Class V, other)

(11) Civilian

(12) Other

3. Select a means of communication; for example, radio, wire, cable, or messenger.

Note: Consider the communications means available and the information's potential significance to your mission. Radio is fast and mobile; yet, normally it is the least secure of the three communications means available at tactical units. Wire is more secure but it is subject to wiretapping and requires more time, personnel and equipment to install. Messenger is very secure but requires more delivery time and is limited by weather, terrain, and threat action.

4. Transmit the message to the receiving authority.

a. If using a messenger, provide the messenger with explicit reporting instructions and a message, preferably written, which is clear, complete, and concise.

b. If using radio, use proper radio/telephone procedures according to unit SOPs. Use the radio only as needed. The enemy may intercept your transmission, exploit the message information, or locate your transmitter for targeting or jamming.

c. If you encounter jamming or interference on your radio net, within 10 minutes of the incident, transmit a meaconing, intrusion, jamming, and interference (MIJI) feeder report, preferably via messenger, wire, or cable to your net control station. Your Signal Operating Instructions (SOI) contains the MIJI format.

Evaluation Preparation: *Setup*: Simulate a situation that requires Soldiers to immediately report information of critical tactical value. You may need two to four personnel (dressed in aggressor uniforms or local attire if available) where they are observable with the naked eye (or binoculars if available). Direct the personnel to perform some type of activity that meets the information requirements. Provide the Soldier with a 1:50,000 scale topographic map of the test area. Provide paper and a pen or pencil for the Soldier to take notes and prepare the report. If you require the Soldier to radio the report to someone else, provide two radios and SOI. Accompany the Soldier being tested to a location where the Soldier can observe the threat.

Brief Soldier: Tell the Soldier he/she is–

Performing an offensive or defensive mission.

Patrolling in a stability or defense support of civil authorities operation.

Manning a checkpoint or roadblock.

Occupying an observation post.

Passing through an area in a convoy.

Instruct the Soldier to report the activity observed, weather factors, and any significant military aspects of the terrain. Once the Soldier completes the report, have the Soldier select a means of transmitting the report to the receiving authority.

Performance Measures	GO	NO GO
1 Identified		
2 Drafted a message in SALUTE format identifying		
3 Selected a means of communication.		
4 Transmitted the message to the receiving authority within 5 minutes of the observation.		

Evaluation Guidance: Refer to chapter 1, paragraph 1-9e, (1) and (2).

References:
Required: TC 3-21.75, FM 6-99
Related: FM 2-91.6

071-COM-0815

Practice Noise, Light, and Litter Discipline

Conditions: You are member of a mounted or dismounted element conducting a tactical mission and have been directed to comply with noise, light and litter discipline. Enemy elements are in your area of operation.

Standards: Prevent enemy from locating your element by exercising noise, light, and litter discipline at all times.

Special Condition: None

Special Standards: None

Special Equipment:

Cue: None

Note: None

Performance Steps

1. Exercise noise discipline.
   a. Avoid all unnecessary vehicular and foot movement.
   b. Secure (with tape or other materials) metal parts (for example, weapon slings, canteen cups, identification [ID] tags) to prevent them from making noise during movement.
   Note: Do not obstruct the moving parts of weapons or vehicles.
   c. Avoid all unnecessary talk.
   d. Use radio only when necessary.
   e. Set radio volume low so that only you can hear.
   f. Use visual techniques to communicate.
2. Exercise light discipline.
   a. Do not smoke.
   Note: The smoking of cigarettes, cigars, etc., can be seen and smelled by the enemy.
   b. Conceal flashlights and other light sources so that the light is filtered (for example, under a poncho).
   c. Cover or blacken anything that reflects light (for example, metal surfaces, vehicles, glass).
   d. Conceal vehicles and equipment with available natural camouflage.
3. Exercise litter discipline.
   a. Establish a litter collection point (empty food containers, empty ammunition cans or boxes, old camouflage) when occupying a position.
   b. Verify all litter has been collected in preparation to leaving a position.
   c. Take all litter with you when leaving a position.

Evaluation Preparation:

Setup: Provide the Soldier with the equipment and or materials described in the conditions statement.

Brief Soldier: Tell the Soldier what is expected of him by reviewing the task standards. Stress to the Soldier the importance of observing all cautions, warnings, and dangers to avoid injury to personnel and, if applicable, damage to equipment.

Performance Measures	GO	NO GO
1 Exercised noise discipline.	_____	_____
2 Exercised light discipline.	_____	_____
3 Exercised litter discipline.	_____	_____

Evaluation Guidance: Score the Soldier GO if all performance measures are passed. Score the Soldier NO-GO if any performance measure is failed. If the Soldier scores a NO-GO, show the Soldier what was done wrong and how to do it correctly.

References
Required:
Related: FM 22-6, TC 3-21.75

071-COM-0801

Challenge Persons Entering Your Area

Conditions: You are a member of a squad or team providing security for your unit in a field environment. You have your assigned weapon, individual protective equipment, and the current challenge and password. An unknown person or persons is approaching your area.

Standards: Detect and challenge all approaching personnel; prevent compromise of password; allow personnel positively identified as friendly to pass; and disarm, detain and report personnel not positively identified.

Special Condition: None

Special Standards: None

Special Equipment:

Cue: None

*Note:*None

Performance Steps

1. Detect all personnel entering your area.
2. Challenge an individual that enters your area.

    a. Cover the individual with your weapon without disclosing your position.

    b. Command the individual to "HALT" before they are close enough to pose a threat.

Note: Commands and questions must be loud enough to be heard by the individual but not loud enough that others outside of the immediate area can hear. Commands should be repeated as necessary.

    c. Ask "WHO IS THERE?" just loud enough for the individual to hear.

Note: The individual should reply with an answer that best describes them, example "Sergeant Jones".

    d. Order the individual to "ADVANCE TO BE RECOGNIZED".

    e. Continue to keep individual covered without exposing yourself.

    f. Order the individual to "HALT" when they are is within 2 to 3 meters from your position.

Note: The individual should be halted at a location that provides protection to you and prevents them from escaping if they are deemed unfriendly.

    g. Issue the challenge in a low voice.

Note: The challenge should only be heard by the individual challenged to prevent all others from overhearing. You may also ask the individual questions that only a friendly person should be able answer correctly.

    h. Determine if the individual is friendly based upon their return of the correct password and your own situational awareness.

        (1) Allow the individual to pass if the individual returns the correct password and you are convinced the individual is friendly.

        (2) Detain an individual if they return an incorrect password or cannot be positively identified as friendly.

            (a) Direct the individual to disarm.

            (b) Notify your chain of command.

            (c) Await instruction from your command.

3. Challenge a group that enters your area.

Note: These procedure and precautions are similar to those for challenging a single person.

    a. Cover the group with your weapon without disclosing your position.

    b. Order the group to halt before they are close enough to pose a threat to you.

    c. Command "WHO IS THERE?" just loud enough to be heard by the group.

    d. Wait for reply from group.

Note: Reply should clearly identify group, example "Friendly Patrol".

    e. Order the leader of the group to "ADVANCE TO BE RECOGNIZED".

    f. Continue to keep group leader covered without exposing yourself.

    g. Order the group leader to "HALT" when the individual is within 2 to 3 meters from your position.

Note: The group leader should be halted at a location that provides protection to you and prevents the leader from escaping if they are deemed unfriendly.

h. Issue the challenge to only the group leader.

Note: You may also ask questions that only a friendly person should be able to answer correctly.

i. Determine if the group leader is friendly based upon their return of the correct password and your own situational awareness.

(1) Direct the group leader to remain with you to assist in identifying group members, if you determine he/she is friendly.

(a) Direct the group leader to vouch for or positively identify each member of his group as they pass to your flank.

(b) Detain any individual in the group not recognized by the group leader by disarming them, and having them wait until your chain of command provides additional instructions.

(2) Detain the group leader, if not positively identified as friendly.

(a) Direct the individual to disarm.

(b) Direct him/her to inform their group to wait.

(c) Notify your chain of command.

(d) Await instruction from your chain of command.

Evaluation Preparation:

Setup: Provide the Soldier with the equipment and or materials described in the conditions statement.

Brief Soldier: Tell the Soldier what is expected of him by reviewing the task standards. Stress to the Soldier the importance of observing all cautions, warnings, and dangers to avoid injury to personnel and, if applicable, damage to equipment.

Performance Measures	GO	NO GO
1 Detected all personnel entering your area.	_____	_____
2 Challenged all individuals that entered your area.	_____	_____
3 Challenged all groups that entered your area.	_____	_____

Evaluation Guidance: Score the Soldier GO if all performance measures are passed. Score the Soldier NO-GO if any performance measure is failed. If the Soldier scores a NO-GO, show the Soldier what was done wrong and how to do it correctly.

References:
Required:
Related: FM 22-6

071-COM-0512

Perform Hand-to-Hand Combat

Conditions: You are a member of a dismounted squad conducting operations and you have encountered an unarmed adversary. You may be equipped with personnel protective equipment (PPE).

Standards: Dominate the enemy using the basic fighting strategy by achieving the clinch, gaining a dominant position and finishing the fight.

Special Condition: None

Special Standards: None

Special Equipment:

Cue: None

*Note:*This task is only a basic introduction to combatives.

Performance Steps

1. Close the distance.
Note: Controlling a standup fight means controlling the range between fighters. The untrained fighter is primarily dangerous at punching range. The goal is to avoid that range. Even if you are the superior striker, the most dangerous thing you can do is to spend time at the range where the enemy has the highest probability of victory.
    a. Achieve the clinch.
       (1) Face your opponent, and assume the Fighting Stance just outside of kicking range.
       (2) Tuck your chin, and use your arms to cover your head while aggressively closing the distance.
       (3) Drive your head into your opponent's chest.
       (4) Move your cupped hands to your opponent's biceps.
       (5) Aggressively fight for one of the dominant clinch positions.
    b. Achieve the Modified Seatbelt Clinch. (Figure 071-COM-0512-1)

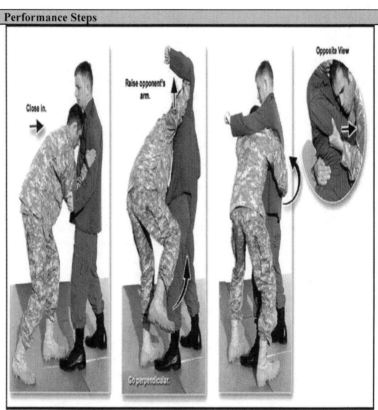

Figure 071-COM-0512-1. Modified Seatbelt Clinch.

    (1) Raise one of his arms.
    (2) Move yourself perpendicular to your opponent.
    (3) Reach around your opponent's waist to grab his opposite-side hip.
    (4) Pull his arm into your chest with your other arm.
    (5) Control his arm at the triceps.
  c. Achieve the Double Under-hooks Clinch. (Figure 071-COM-0512-2)

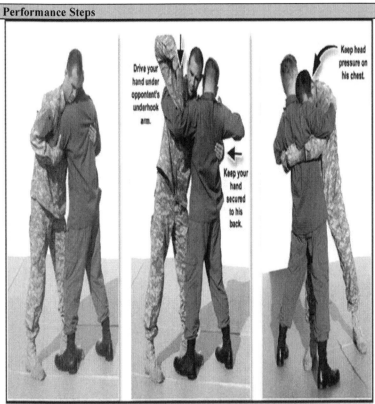

Figure 071-COM-0512-2. Double Underhooks Clinch.

(1) Drive your overhook hand (with a knife edge) under your opponent's underhook arm.

(2) Clasp your hands in a Wrestler's Grip behind your opponent, while keeping head pressure on his chest.

d. Achieve the Rear Clinch. (Figure 071-COM-0512-3)

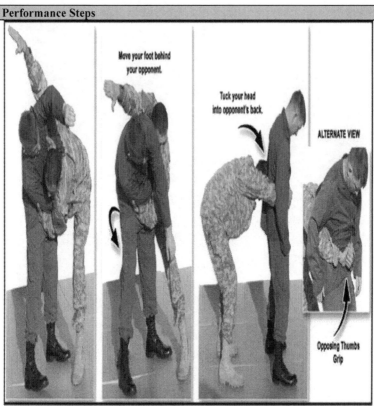

Figure 071-COM-0512-3. The Rear Clinch.

(1) Step behind your opponent.

(2) Clasp your hands around your opponent's waist in an Opposing Thumbs Grip.

(3) Place your forehead in the small of his back to avoid strikes.

Note: From this secure position, you can attempt to take the opponent down.

   2. Gain dominant position.

      a. Achieve the Rear Mount. (Figure 071-COM-0512-4)

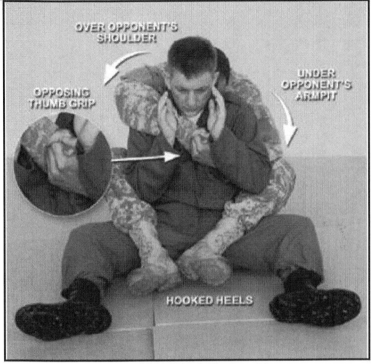

Figure 071-COM-0512-4. The Rear Mount

     (1) Place one arm under your opponent's armpit and the other over his opposite shoulder.

     (2) Clasp your hands in an Opposing Thumbs Grip.

WARNING

When in the Rear Mount, DO NOT cross your feet; this would provide the opponent an opportunity for an ankle break.

---

     (3) Wrap both legs around your opponent, with your heels hooked inside his legs.

Note: Keep your head tucked to avoid headbutts.

     b. Achieve the Mount. (Figure 071-COM-0512-5)

Note: The Mount allows the fighter to strike the opponent with punches, while restricting the opponent's ability to deliver effective return punches. The Mount also provides the leverage to attack the opponent's upper body with chokes and joint attacks.

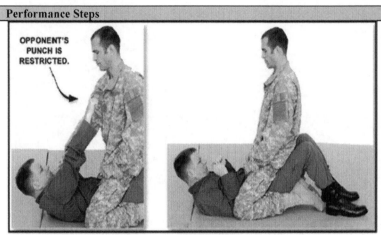

OPPONENT'S
PUNCH IS
RESTRICTED.

Figure 071-COM-0512-5. The Mount.

(1) Position your knees as high as possible toward the opponent's armpits.

(2) Place your toes in line with or inside of your ankles to avoid injuring your ankles when your opponent attempts to roll you over.

    c. Achieve the Guard. (Figure 071-COM-0512-6)

Note: A fighter never wants to be under his opponent; the Guard enables him to defend himself and transition off of his back into a more advantageous position. The Guard allows the bottom fighter to exercise a certain amount of control over the range by pushing out or pulling in his opponent with his legs and hips. With skill, the bottom fighter can defend against strikes and even apply joint locks and chokes.

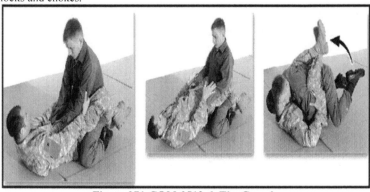

Figure 071-COM-0512-6. The Guard.

(1) Control opponent's arms at the elbows.

(2) Lock your ankles around opponent's torso.

    d. Achieve Side Control. (Figure 071-COM-0512-7)

Figure 071-COM-0512-7. Side Control.

(1) Keep the leg closest to your opponent's head straight.

(2) Bend the other leg so that the knee is near your opponent's hip.

(3) Keep your head turned away to avoid knee strikes.

(4) Place your elbow on the ground in the notch created by the opponent's head and shoulder.

(5) Position your other hand palm down on the ground under the opponent's near-side hip.

3. Finish the Fight.

Note: When dominant body position has been achieved, the fighter can begin attempts to finish the fight secure in the knowledge that if an attempt fails, as long as he maintains dominant position, he may simply try again.

a. Achieve the Rear Naked Choke.(Figure 071-COM-0512-8)

Note: The Rear Naked Choke slows the flow of blood in the carotid arteries, which can eventually cause your opponent to be rendered unconscious for a short period of time.

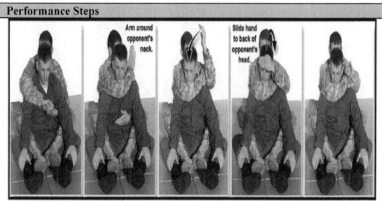

Figure 071-COM-0512-8. Rear Naked Choke.

(1) Place your bicep along one side of your opponent's neck; take your forearm and roll it around to the other side of neck, his chin will line up with your elbow.

(2) Tighten choke up and grab your shoulder or arm.

(3) Place your opposite hand behind the head as if your combing his hair back.

(4) Tuck your head in to avoid getting hit.

(5) Roll your shoulders back, push chest forward and finish the choke.

b. Achieve the Cross Collar Choke from the Mount and Guard. (Figure 071-COM-0512-9 and 10)

Note: The Cross-Collar Choke is a blood choke that can only be employed when your opponent is wearing a durable shirt. This choke should be performed from either the Mount or Guard.

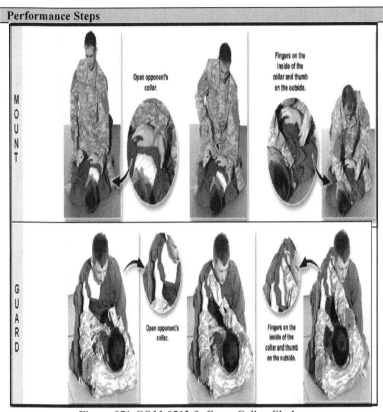

Figure 071-COM-0512-9. Cross Collar Choke.

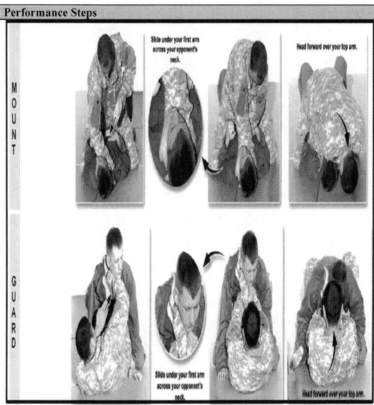

Figure 071-COM-0512-10. Cross Collar Choke Continued.

(1) Open your opponent's same-side collar With your non-dominant hand.

(2) Reach across your body, and insert your dominant hand into the collar you just opened.

(3) Relax the dominant hand, and reach all the way behind your opponent's neck.

(4) Grasp his collar with your fingers on the inside and your thumb on the outside.

(5) Release the grip of your non-dominant hand, and move your dominant-side forearm across your opponent's neck under the first arm, clearing his chin.

(6) Reach all the way back untilyour dominant hand meets the other hand using the same grip.

(7) Turn your wrists so that your palms face you, and pull your opponent into you.

(8) Expand your chest,and pinch your shoulders together.

(9) Bring your elbows to your hips to finish the choke.

c. Achieve the Bent Arm Bar from the Mount and Side Control. (Figure 071-COM-0512-11)

Note: The Bent Arm Bar is a joint lock that attacks the shoulder girdle. This technique can be employed from either the Mount or Side Control.

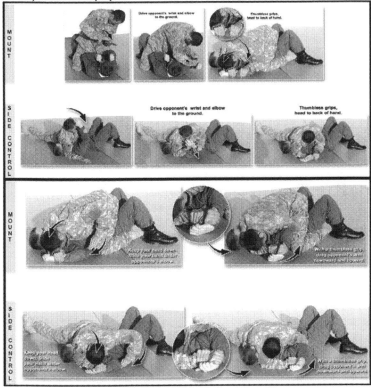

Figure 071-COM-0512-11. Bent Arm Bar.

(1) Drive your opponent's wrist and elbow to the ground with thumbless grip.

(2) Move your elbow to the notch created by your opponent's neck and shoulder.

(3) Keep your head on the back of your hand to protect your face from strikes.

(4) Place your other hand under his elbow.

(5) Grab your own wrist with a Thumbless Grip.

(6) Drag the back of your opponent's hand toward his waistline.

(7) Lift his elbow, and dislocate his shoulder.

d. Achieve the Straight Arm Bar from the Mount. (Figure 071-COM-0512-12)

Note: The Straight Arm Bar is a joint lock designed to damage the elbow. While this exercise outlines a Straight Arm Bar performed from the Mount, this technique can be performed from any dominant position.

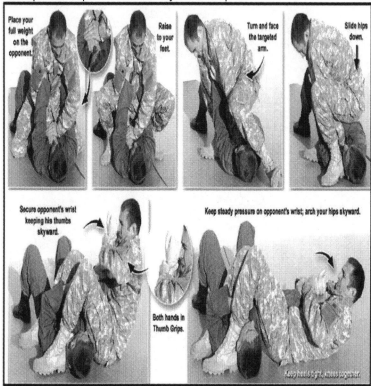

Figure 071-COM-0512-12. Straight Arm Bar from the Mount.

(1) Decide which arm you wish to attack.

(2) Isolate that arm by placing your opposite-side hand in the middle of your opponent's chest, between his arms.

(3) Target the unaffected arm and press down to prevent your opponent from getting off the flat of his back.

(4) Loop your same-side arm around the targeted arm and place that hand in the middle of your opponent's chest, applying greater pressure.

(5) Place all of your weight on your opponent's chest and raise to your feet in a very low squat.

(6) Turn your body 90 degrees to face the targeted arm.

(7) Bring the foot nearest to your opponent's head around his face, and plant it in the crook of his neck on the opposite side of the targeted arm.

(8) Slide your hips down the targeted arm, keeping your buttocks tight to your opponent's shoulder.

(9) Secure your opponent's wrist with both of your hands in Thumb Grips.

(10) Keep his thumb pointed skyward to achieve the correct angle.

(11) Pull your heels tight to your buttocks, and pinch your knees together tightly with the upper arm trapped between your knees, not resting on your groin.

(12) Apply steady pressure by trapping your opponent's wrist on your chest, and arching your hips skyward.

e. Achieve the Straight Arm Bar from the Guard. (Figure 071-COM-0512-13)

Note: Fighting from your back can be very dangerous. When your opponent attempts to strike and apply chokes from within your Guard, use the Straight Arm Bar from the Guard, a joint lock designed to damage the elbow.

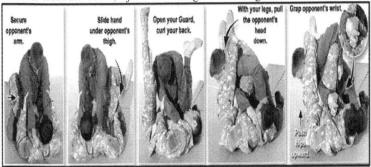

Figure 071-COM-0512-13. Straight Arm Bar from the Guard.

(1) Secure the arm at or above the elbow when your opponent presents a straight arm.

(2) Hold your opponent's elbow for the remainder of the move.

(3) Insert your other hand under the opponent's thigh on the side opposite the targeted arm.

Note: The hand should be palm up.

(4) Open your Guard, and bring your legs up, while curling your back to limit the friction.

(5) Contort your body by pulling with the hand that is on the back of your opponent's thigh.

(6) Bring your head to his knee.

(7) Place your leg over his head.

(8) Grab your opponent, and pull him down by pulling your heels to your buttocks and pinching your knees together with your leg.

(9) Move the hand that was behind your opponent's thigh to grasp the wrist that you secured at the elbow with a Thumb Grip.

(10) Curl your calf downward and push up with your hips to break your opponent's arm.

f. Achieve the Guillotine Choke. (Figure 071-COM-0512-14 and Figure 071-COM-0512-15)

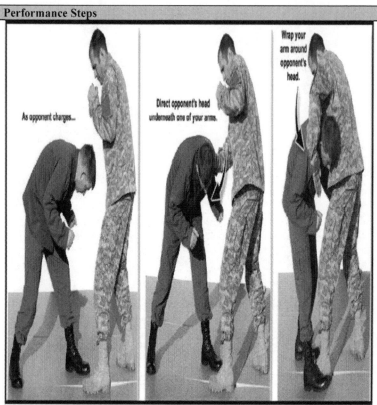

Figure 071-COM-0512-14. Guillotine Choke.

Figure 071-COM-0512-15. Guillotine Choke Continued.

(1) Direct your opponent's head underneath one of your arms, and take a step back when your opponent charges your legs.

(2) Wrap your arm around your opponent's head and under his neck.

(3) Grasp the first hand where a watch would be with your other hand, ensuring that you have not reached around your opponent's arm.

(4) Cinch the choke by bringing your arm further around your opponent's head, improving your grip.

(5) Cinch up the choke and sit down to place him in your guard.

Note: Your palm should be facing your own chest.

(6) Sit Down.

(7) Place your opponent within your Guard.

(8) Finish the choke by pulling with your arms and pushing with your legs.

Evaluation Preparation:

Setup: Provide the Soldier with the equipment and or materials described in the conditions statement.

Brief Soldier: Tell the Soldier what is expected of him by reviewing the task standards. Stress to the Soldier the importance of observing all cautions, warnings, and dangers to avoid injury to personnel and, if applicable, damage to equipment.

Performance Measures	GO	NO GO
1 Achieved the Clinch.	____	____
2 Gained a dominant position.	____	____
3 Finished the Fight.	____	____

Evaluation Guidance: Score the Soldier GO if all performance measures are passed. Score the Soldier NO-GO if any performance measure is failed. If the Soldier scores a NO-GO, show the Soldier what was done wrong and how to do it correctly.

References:
Required:
Related: TC 3-25.150

071-COM-4408

Construct Individual Fighting Positions

Conditions: You are a member of a squad that has just occupied a defense position and you have been directed to construct an individual fighting position. You have your assigned weapon(s) (M249 machine gun, M240B machine gun, M16-series rifle, M4- series carbine, and/or a shoulder launched missile), a blank DA Form 5517-R Standard Range Card, personal protective equipment, construction material, and camouflage material. You have been given your sectors of fire. Some iterations of this task should be performed in MOPP 4.

Standards: Construct a fight position based on leadership direction and type of weapon(s) assigned. Ensure fighting position provides frontal, side, rear, and overhead cover (OHC), as required. Prepare a range card for the position.

Special Condition: None

Safety Risk: Low

MopP 4: Sometimes

Cue: None

Note: A fighting position provides cover from fire and concealment from observation while allowing you to engage the enemy with your weapon. There are two types of fighting position: hasty and deliberate. The type of fighting position you construct is dependent on: time available, equipment available, and the required level of protection required.

If assigned an M4 rather than an M16-series weapon, add 7 inches (18 centimeters). The length of two M16s is equal to two and a half M4s. The widths of all the fighting positions are only an approximate distance and based on the individual Soldier's equipment.

OHC can be built up or down, this task covers built up OHC. Built-up OHC is constructed on top of the parapets up to 18 inches (46 centimeters) and provides for maximum room inside the fighting position and adequate space between the end walls of the fighting position and the OHC. Built-down OHC is constructed at or below ground level and should not exceed 12 inches (30 centimeters) above ground. This lowers the profile of the fighting position, which aids in avoiding detection. However, it restricts the fighting space between the end walls of the fighting position and the OHC. To account for this restricted space the width of the fighting position should be extended to three M16 lengths.

Performance Steps

1. Construct a hasty fighting position

Note: A hasty fighting position should give frontal cover from enemy direct fire but allow firing to the front and the oblique. Hasty positions are used if: there is little time for preparation, there is no requirement for a deliberate defensive position (such as a pause during movement) or you have just occupied the position. A hasty fighting position uses whatever cover is available. The position can be developed into a deliberate position, if in a suitable location.

    a. Construct a shell crater.

      (1). Lie prone in the depression.

      (2). Orient your position so you are oblique to enemy fire.

    b. Construct a skirmisher's trench.

      (1). Physical with firearms used.

Note: A skirmisher's trench is used for immediate shelter from enemy fire when there are no defilade firing positions available. In all but the hardest ground, you can use this technique to quickly form a shallow, body-length pit

      (1) Lie prone or on your side.

      (2). Report the situation immediately to the section or team leader.

      (3). Scrape the soil underneath or beside you with an entrenching tool.

      (4). Pile the soil in a low parapet between yourself and the enemy

    c. Construct a prone fighting position (Figure 071-COM-4408-1).

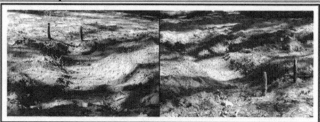

Figure 071-COM-4408-1.
Example of a prone fighting position (Hasty)

(1). Construct a crater or skirmisher's trench fighting position.

(2). Scrape additional soil from your position to a depth of about 18 inches (46 centimeters).

(3). Build cover around the edge of the position by using the dirt dug from the hole.

2. Construct a deliberate fighting position.

a. Construct a one-man fighting position.

Note: Except for its size, a one-man position is built the same way as a two-man fighting position. The hole of a one-man position is only large enough for you and your equipment. It does not have the security of a two-person position; therefore, it must allow you to shoot to the front or oblique from behind frontal cover.

b. Construct a two-man fighting position. (Figure 071-COM-4408-2).

Note: A two-man fighting position is preferred over the one-man fighting position as it allows more flexibility and better security. A two-man fighting position is constructed in four stages with the chain of command normally inspecting and providing additional guidance between each phase.

Figure 071-COM-4408-2.
Two-man fighting position with OHC.

(1). Construct stage 1 of a two-man fighting position. (Figure 071-COM-4408-3).

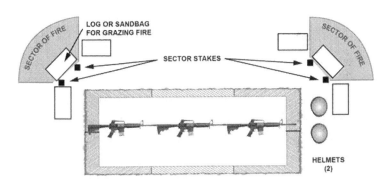

Figure 071-COM-4408-3.
Two-man fighting position - stage 1 (top view).

(a). Identify sector(s) of fire (at least primary and possibly secondary).

(b). Check fields of fire from the prone position.

(c). Emplace sector stakes (right and left) to define your sectors of fire.

Note: The sector stakes must be sturdy and stick out of the ground at least 18 inches (46 centimeters); this will prevent your weapon from being pointed out of your sector.

    (d). Emplace aiming and limiting stakes as needed.

Note: Aiming and limiting stakes help you fire into dangerous approaches at night and at other times when visibility is poor. Forked tree limbs about 12 inches (30 centimeters) long make good stakes. One stake (possibly sandbags) is placed near the edge of the hole to rest the stock of your rifle on. The other stake is placed forward of the rear (first) stake/sandbag toward each dangerous approach. The forward stakes are used to hold the rifle barrel.

    (e). Emplace grazing fire logs or sandbags to achieve grazing fire 1 meter above ground level.

    (f). Scoop out elbow holes to keep your elbows from moving around when you fire.

    (g). Trace position outline.

Note: The length of two M16s is equal to two and a half M4s. The widths of all the fighting positions are only an approximate distance and based on the individual Soldier's equipment.

    (h). Clear primary and secondary (if applicable) fields of fire

    (2). Construct stage 2 of a two-man fighting position. (Figure 071-COM-4408-4).

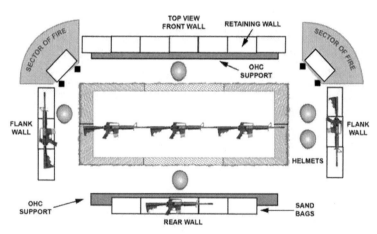

Figure 071-COM-4408-4
Two-man fighting position - stage 2 (top view).

    (a). Emplace OHC supports to front and rear of position, at least 12 inches (30 centimeters) from the edge of the position outline.

Note: 12 inches (30 centimeters) is about 1-helmet length

If you plan to use logs or cut timber, secure them in place with strong stakes from 2 to 3 inches (5 to 7 centimeters) in diameter and 18 inches (46 centimeters) long. Short U-shaped pickets will work.

   (b). Construct parapet retaining walls.

    1. Construct the front retaining wall at least 10 inches (25 centimeters) high, two filled sandbags deep, and equal length of the fighting position.

    2. Construct rear retaining wall--At least 10 inches (25 centimeters) high, and one M16 long.

    3 Construct flank retaining walls--At least 10 inches (25 centimeters) high, and equal width of the fighting position.

   (c). Remove the top layer of dirt from the hole.

    1. Set aside grass or foliage for camouflage

    2. Use excavated soil to fill sandbags

  (3). Construct stage 3 of a two-man fighting position

   (a). Dig position with vertical walls to a maximum depth of armpit deep (if soil conditions permit). (Figure 071-COM-4408-5)

Note: If the walls of the position are unstable, due to soil properties, you can use revetments and/or slope the walls. Plywood or sheeting material and pickets can be used to revet walls. For sloped walls you would first dig a vertical hole and then slope the walls at 1:4 ratio (move 12 inches [30 centimeters] horizontally for each 4 feet [1.22 meters] vertically).

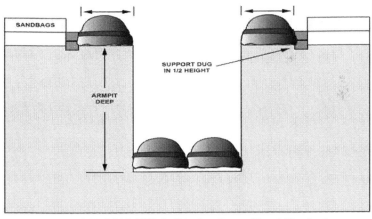

Figure 071-COM-4408-5. Digging the position (side view)

  (b). Use excavated soil from hole to fill parapets in the order of front, flanks, and rear.

   (c). Verify you can cover the entire sector of fire from this position

   (d). Dig two grenade sumps in the floor one on each end.

Note: Grenade sumps are as wide as the entrenching tool blade; at least as deep as an entrenching tool and as long as the position floor is wide.

(e). Slope the floor toward the grenade sumps.

(f). Dig a storage compartment in the bottom of the back wall; the size of the compartment depends on the amount of equipment and ammunition to be stored.

(g). Install revetments, if required, to prevent wall collapse/cave-in.

(h). Emplace standard length stringers for OHC (Figure 071-COM-4408-6).

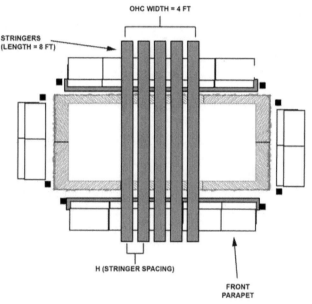

Figure 071-COM-4408-6
Placement of stringers for OHC

(4). Construct stage 4 of a two-man fighting position.

(a). Install OHC. (Figure 071-COM-4408-7).

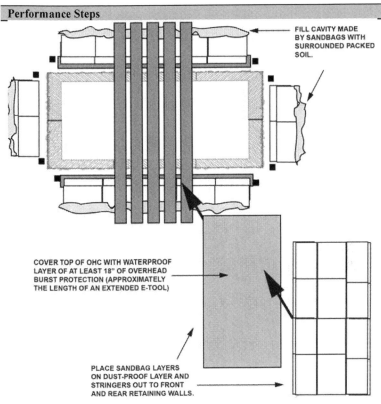

FILL CAVITY MADE BY SANDBAGS WITH SURROUNDED PACKED SOIL.

COVER TOP OF OHC WITH WATERPROOF LAYER OF AT LEAST 18" OF OVERHEAD BURST PROTECTION (APPROXIMATELY THE LENGTH OF AN EXTENDED E-TOOL)

PLACE SANDBAG LAYERS ON DUST-PROOF LAYER AND STRINGERS OUT TO FRONT AND REAR RETAINING WALLS.

Figure 071-COM-4408-7
Installation of OHC

    1. Emplace dustproof layer.

Note: Plywood, sheeting mats can be used as a dustproof layer (could be boxes, plastic panel, or interlocked U-shaped pickets). A standard dustproof layer is 4'x4' sheets of ¾-inch plywood centered over dug position

    2. Nail plywood dustproof layer to stringers, if required

    3. Emplace at least 18 inches (46 centimeters) of filled sandbags for overhead burst protection (*Note:* At a minimum four layers.) the sandbags must cover the area between the front and rear retaining wall.

    4. Use plastic or a poncho for waterproofing layer.

    5. Fill center cavity with soil from dug hold and surrounding soil.

    (b). Camouflage the fighting position.

        1. Mold the OHC and parapets to blend with the surrounding terrain

        2. Camouflage the position with natural materials that do not have to be replaced

Note: Rocks, logs, live bushes, grass, and other available materials can be used to make the position blend with surroundings, or camouflage screen systems

       3. Ensure the position cannot be seen within 115 feet (35 meters).

   3. Construct a machine gun fighting position. (Figure 071-COM-4408-8 and Figure 071-COM-4408-9)

071-COM-4408-8.
Machine gun fighting position with OHC.

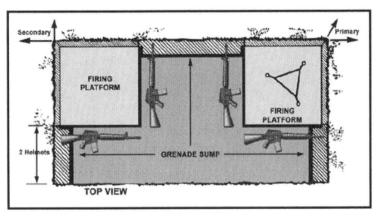

071-COM-4408-9.

Machine gun fighting position (top view).

a. Construct stage 1 of a machine gun fighting position.

(1). Establish sectors (primary and secondary) of fire

(a) Check fields of fire from the prone position.

(b) Assign sector of fire (primary and secondary) and final protective line (FPL) or principal direction of fire (PDF).

(c) Emplace aiming stakes.

(d) Decide whether to build OHC up or down, based on potential enemy observation of position.

(2). Mark the outline of the position.

(a) Trace position outline to include location of two distinct firing platforms.

(b) Mark position of the tripod legs where the gun can be laid on the FPL or PDF.

(3). Clear primary and secondary fields of fire.

b. Construct Stage 2 of a machine gun fighting position.

(1). Dig firing platforms 6 to 8 inches (15 to 20 centimeters) deep and one M16 in length and width.

(2). Emplace the OHC supports to front and rear of the position.

Note: The supports are placed the same as for a two-man fighting position.

(3). Construct the parapet retaining walls.

Note: The parapet retaining walls are constructed the same as for a two-man fighting position.

(4). Position the machine gun to cover primary sector of fire.

c. Construct stage 3 of a machine gun fighting position.

(1) Dig position and build parapets.

(a) Dig the position to a maximum armpit depth around the firing platform.

(b) Use soil from hole to fill parapets in order of front, flanks, and rear.

(c) Dig grenade sumps and slope floor toward them.

(d) Install revetment if needed.

Note: Follow same steps as for two-man fighting position.

(2) Place stringers for OHC.

Note: Stringers are placed the same way as for a two-man position.

d. Construct stage 4 of a machine gun fighting position.

(1) Install OHC.

Note: Build the OHC the same as you would for a two-man fighting position.

(2) Install camouflage.

(a) Use surrounding topsoil and camouflage screen systems.

(b) Ensure position cannot be seen within 115 feet (35 meters).

(c) Use soil from hole to fill sandbags and OHC cavity, or to spread around and blend position in with surrounding ground.

4. Construct a shoulder launched missile fighting position.

a. Construct an M136 fighting position

Note: An M136 fighting position is a standard two-man fighting position that includes basic considerations for firing shoulder launched missile. The shoulder launched missile is fired from a modified standing position by leaning against the rear wall of the fighting position and ensuring the rear of the weapon extends beyond the rear of the fighting position.

    (1) Construct stage 1.

Note: Only additional consideration is identifying the backblast area to ensure it is kept cleared. Leaders must ensure that shoulder launched missiles are positioned so that the backblast misses other fighting positions.

    (2) Construct stage 2.

Note: Only additional consideration is the rear parapet does not block the backblast area.

    (3) Construct stage 3.

Note: No additional considerations.

    (4) Construct stage 4.

Note: Only additional consideration is ensuring any camouflage in the backblast area is secure and not easily combustible.

    b. Construct a standard Javelin fighting position with OHC.

Note: The standard Javelin fighting position has cover to protect you from direct and indirect fires. The position is prepared the same as the two-man fighting position with two additional steps. See Figure 071-COM-4408-10.

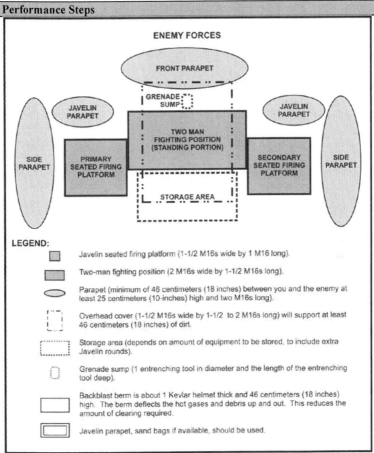

Figure 071-COM-4408-10
Standard Javelin firing position

(1) Extend and slope the back wall of the position rearward to serve as a storage area.

(2) Extend the front and side parapets twice the length as the dimensions of the two-man fighting position with the Javelin's primary and secondary seated firing platforms added to both sides.

5. Prepare a DA Form 5517-R, Standard Range Card for the fighting position.

Note: A range card is comprised of sectors of fire, principal direction of fire, final protective live, and dead space.

a. Orient the card so both the primary and secondary sectors of fire can fit on it.

b. Draw a rough sketch of the terrain to the front of your position.

Note: Include any prominent natural and man-made features that could be likely targets.

 c. Draw your position at the bottom of the sketch.

Note: Do not put in the weapon symbol at this time.

 d. Fill in the marginal data.

  (1) Gun number or squad.

  (2) Platoon, company and date.

  (3) Magnetic north arrow.

 e. Sketch in the magnetic north arrow on the card with its base starting at the top of the marginal data section.

 f. Using your compass, determine the azimuth in degrees from the terrain feature to the gun position.

 g. Determine the distance between the gun and the feature by pacing or plotting the distance on a map.

 h. Sketch in the terrain feature on the card in the lower left or right hand corner.

 i. Connect the sketch of the position and the terrain feature with a barbed line from the feature to the gun.

 j. Write in the distance in meters.

 k. Add final protective fires to your range card.

  (1) Sketch in the limits of the primary sector of fire as assigned by your leader.

  (2) Sketch in the FPL line on your sector limit as assigned.

  (3) Determine dead space on the final protective line by having your assistant gunner walk the final protective line.

  (4) Watch him walk down the line and mark spaces that cannot be grazed.

  (5) Sketch dead space by showing a break in the symbol for an FPL, and write in the range to the beginning and end of the dead space.

  (6) Label all targets in your primary sector in order of priority.

 l. Prepare range card when assigned a PDF instead of an FPL.

  (1) Sketch in the limits of the primary sector of fire as assigned by your leader.

Note: Sector should not exceed 875 mils, the maximum traverse of the tripod-mounted machine gun.

  (2) Sketch in the symbol for an automatic weapon oriented on the most dangerous target within your sector Note: The PDF will be target number one in your sector. All other targets will be numbered in priority.

  (3) Sketch in your secondary sector of fire.

Note: The secondary sector is drawn using a broken line.

  (4) Label targets within the secondary sector with the range in meters from your gun to each target.

Note: When necessary the bipod is used to engage targets in your secondary sector.

  (5) Sketch in aiming stakes, if used.

Evaluation Preparation:

Setup: Provide the Soldier with the equipment and or materials described in the conditions statement.

Brief Soldier: Tell the Soldier what is expected of him by reviewing the task standards. Stress to the Soldier the importance of observing all cautions, warnings, and dangers to avoid injury to personnel and, if applicable, damage to equipment.

Performance Measures	GO	NO GO
1 Constructed a hasty fighting position.	_____	_____
2 Constructed a deliberate fighting position.	_____	_____
3 Constructed a machine gun fighting position.	_____	_____
4 Constructed a shoulder launched missile fighting position.	_____	_____
5 Prepared a DA Form 5517-R, Standard Range Card for the fighting position.	_____	_____

Evaluation Guidance: Score the Soldier GO if all performance measures are passed. Score the Soldier NO-GO if any performance measure is failed. If the Soldier scores a NO-GO, show the Soldier what was done wrong and how to do it correctly.

References:
Required: DA FORM 5517-R, TC 3-21.75, TM 3-23.25
Related:

052-COM-1361

Camouflage Yourself and Your Individual Equipment

Conditions: Given an individual weapon, grass, bushes, and trees, pieces of the Lightweight Camouflage Screen System (LCSS), skin paint, and charcoal and/or mud. Some iterations of this task should be performed in MOPP 4.

Standards: Camouflage yourself and your individual equipment to prevent detection by visual, near-infrared, infrared, ultraviolet, radar, acoustic, and radio sensors.

Special Condition: None

Safety Risk: Low

Special Standards: None

Special Equipment:

MOPP 4: Sometimes

Cue: None

Note:

Performance Steps

1. Apply camouflage principles throughout camouflaged operations.
    a. Employ realistic camouflage.
        (1) Employ camouflage material that resembles the background.
        (2) Employ camouflage subtly without overdoing.
    b. Apply camouflaged movement technique.
Note: Movement draws attention, and darkness does not prevent observation. The naked eye and infrared/radar sensors can detect movement.
        (1) Minimize movement.
        (2) Move slowly and smoothly when movement is necessary.
    c. Breakup regular shapes.
        (1) Use natural or artificial materials to breakup shapes, outlines, and equipment.
        (2) Stay in shadows when moving, if possible.
        (3) Disguise or distort the shape of your helmet and your body with natural or artificial materials when conducting operations close to the enemy.

d. Reduce possible shine by covering or removing items that may reflect light.

Note: Examples of items that should be covered and/or removed include: mirrors, eye glasses, watch crystals, plastic map cases, starched uniforms, clear-plastic garbage bags, red-filtered flashlights, goggles worn on top of helmets cigarettes and pipes.

e. Blend colors with the surroundings or, at a minimum, ensure that objects do not contrast with the background (figure 052-COM-1361-1).

Note: Change camouflage, as required, when moving from one area to another. What works well in one location may draw fire in another

Figure 052-COM-1361-1.
Colors Used for Camouflage

f. Employ noise discipline.

2. Camouflage your exposed skin.

Note: Exposed skin reflects light.

a. Cover your skin oils, using paint sticks, even if you have very dark skin.

Note: Paint sticks cover these oils and provide blending with the background.

b. Use the color chart in table 052-COM-1361-1 when applying paint on the face.

Table 052-COM-1361-1. Color Chart.

Camouflage Material	Skin Color Light or Dark	Shine Areas Forehead, Cheekbones, Ears, Nose, and Chin	Shadow Areas Around Eyes, Under Nose, and Under Chin
Loam and Light Green Stick	All troops use in areas with green vegetation	Use loam	Use light green
Sand and Light Green Stick	All troops use in areas lacking green vegetation	Use light green	Use sand
Loam and White Stick	All troops use only in snow covered terrain	Use loam	Use white
Burnt Cork, Bark Charcoal, or Lamp Black	All troops use if camouflage sticks are not available	Use	Do not use
Light – Color Mud	All troops use if camouflage sticks are not available	Do not use	Use

    c. Paint high, shiny areas (forehead, cheekbones, nose, ears, and chin) with a dark color

    d. Paint low, shadow areas (around the eyes, under the nose and under the chin) with a light color.

CAUTION

Mud contains bacteria, some of which is harmful and may cause disease or infection. Mud should be considered as a last resort for field expedient paint.

Expedient paint containing motor oil should be used with extreme caution. Prolonged exposure to motor oil may result in personal injury.

---

    e. Paint exposed skin on the back of the neck, arms, and hands with an irregular pattern.

    3. Camouflage your uniform and helmet.

      a. Roll your sleeves down, and button all buttons.

CAUTION

Soldiers must be aware of local foliage hazards, and possible reactions to poisonous leaves.

.

  b. Attach leaves, grass, small branches, or pieces of LCSS to your uniform and helmet (figure 052-COM-1361-2). These items will distort shapes and blend colors with the natural background
Note: ACUs provide visual and near-infrared camouflage

Figure 052-COM-1361-2.
Camouflaged Helmets.

    c. Wear unstarched ACUs.

Note: Starch counters the infrared properties of the dyes.

    d. Replace excessively faded and worn ACUs because camouflage effectiveness is lost.

Performance Steps

4. Camouflage your personal equipment
   a. Cover or remove shiny items.
   b. Secure items that rattle or make noise when moved or worn.
   c. Breakup the shape of large and bulky equipment using natural items and/or LCSS.
5. Maintain camouflage.
   a. Replace natural camouflage as it dies and loses its effectiveness.
   b. Replace camouflage as it fades.
   c. Replace camouflage to correspond to changing surroundings.

Evaluation Preparation:

Setup: Provide the Soldier with the equipment and or materials described in the conditions statement.

Brief Soldier: Tell the Soldier what is expected of him by reviewing the task standards. Stress to the Soldier the importance of observing all cautions, warnings, and dangers to avoid injury to personnel and, if applicable, damage to equipment.

Performance Measures	GO	NO GO
1 Applied camouflage principles throughout camouflaged operations.	____	____
a. Employed realistic camouflage.	____	____
b. Applied camouflaged movement technique.	____	____
c. Broke-up regular shapes.	____	____
d. Reduced possible shine by covering or removing items that may reflect light.	____	____
e. Blended colors with the surroundings or, at a minimum, ensured that colors so not contrast with the background.	____	____
f. Employed noise discipline.	____	____
2 Camouflaged your exposed skin.		
a. Covered your skin oils, using paint sticks, even if you have very dark skin.	____	____
b. Used the color chart in table 052-COM-1361-1 when applying paint on the face.	____	____
c. Painted high, shiny areas (forehead, cheekbones, nose, ears, and chin) with a dark color.	____	____
d. Painted low, shadow areas (around the eyes, under the nose and under the chin) with a light color.	____	____
e. Painted exposed skin on the back of the neck, arms, and hands with an irregular pattern.	____	____

Performance Measures	GO	NO GO
3 Protected yourself against physical and other hazards.	___	___
a. Rolled your sleeves down, and buttoned all buttons.	___	___
b. Attached leaves, grass, small branches, or pieces of LCSS to your uniform and helmet.	___	___
c. Wore unstarched ACUs.	___	___
d. Replaced excessively faded and worn ACUs because camouflage effectiveness is lost.	___	___
4 Camouflaged your personal equipment.		
a. Covered or removed shiny items.	___	___
b. Secured items that rattle or make noise when moved or worn.	___	___
c. Broke-up the shape of large and bulky equipment using natural items and/or LCSS.	___	___
5 Maintained camouflage.		
a. Replaced natural camouflage as it dies and loses its effectiveness.	___	___
b. Replaced camouflage as it fades.	___	___
c. Replaced camouflage to correspond to changing surroundings.	___	___

Evaluation Guidance: Score the Soldier GO if all performance measures are passed. Score the Soldier NO-GO if any performance measure is failed. If the Soldier scores a NO-GO, show the Soldier what was done wrong and how to do it correctly.

References
Required: ATP 3-37.34, TC 3-21.75
Related:

071-COM-0011

Employ Progressive Levels of Individual Force

> **WARNING**
> During the assessment ensure that biological threats associated with close
> contact/combat are taken into consideration and protective measures are taken
> to prevent exposure.

Conditions: You are a member of a section or team that is securing a critical area or defusing a civil disturbance and you are approached/confronted by one or more hostile civilians. You have your individual weapon, personal protection equipment (PPE), and the rules of engagement (ROE).

Standards: Assess and immediately report threats situations to your leadership. Protect yourself against hazards. Isolate hostile civilians, if required. Control the situation using the least amount of force possible.

Special Condition: None

Special Standards: None

Special Equipment:

Cue: None

Note: The operational environment must be considered at all times during this task. All Army elements must be prepared to enter any environment and perform their missions while simultaneously dealing with a wide range of unexpected threats and other influences. Units must be ready to counter these threats and influences and, at the same time, be prepared to deal with various third-party actors, such as international humanitarian relief agencies, news media, refugees, and civilians on the battlefield. These groups may or may not be hostile to us, but they can potentially affect the unit's ability to accomplish its mission.

Performance Steps

1. Assess the situation by identifying the level of hostile civilian threat.
    a. Verbal.
    b. Physical without weapons (touching, pushing).
    c. Physical with weapons (rocks, clubs, spitting).
    d. Physical with firearms shown.
    e. Physical with firearms used.
2. Report the situation immediately to the section or team leader.

3. Protect yourself against physical and other hazards.

   a. Use full - face shields.

   b. Use double layer latex gloves.

Note: Any exposure incident must be reported to the chain of command.

4. Isolate hostile civilian(s), if required.

   a. Identify hostile group(s) sphere of influence.

   b. Remove the individual with the most influence of the crowd.

   c. Use the 5S's (Search, Silence, Segregate, Safeguard, Speed to the rear).

5. Employ no more force than is necessary to control the situation using graduated response measures.

Note: Soldiers should employ the lowest level of force necessary to address a threat but may use any level, even deadly force, without performing earlier steps, if the circumstances or threat do not allow for the use of graduated levels of force.

   a. Avoid confrontation if possible.

   b. Do not deliberately instigate, threaten, provoke, or bluff.

   c. Speak sternly to the civilian and state the peaceful intent of your mission.

   d. Tell the civilian to "STAND BACK" and warn them that you may have to use force.

   e. If a civilian places his or her hands on your body, brush them back with hand or availble PPE.

   f. If a civilian attempts to inflict bodily harm, use any authorized materials (such as water hoses, chemical gases) to impede movement.

   g. Use your individual weapon, if necessary, as prescribed by the established ROE.

6. Establish and maintain control of the situation.

   a. Comply with the ROE, any host-nation requirements, applicable international treaties and operational agreements.

Note: ROE are directives issued by competent military authority that delineate the circumstances and the limitations under which United States forces will initiate and/or continue combat engagement with other forces encountered. ROE help commanders accomplish the mission by regulating the rules of the use of force. Everyone must understand the ROE and be prepared to execute them properly in every possible confrontation.

   b. Minimize casualties and damage.

   c. Maintain professional demeanor and appearance.

Evaluation Preparation:

Setup: Provide the Soldier with the equipment and or materials described in the conditions statement.

Brief Soldier: Tell the Soldier what is expected of him by reviewing the task standards. Stress to the Soldier the importance of observing all cautions, warnings, and dangers to avoid injury to personnel and, if applicable, damage to equipment.

Performance Measures	GO	NO GO
1 Assessed the situation by identifying the level of hostile civilian threat.	____	____
2 Reported the situation immediately to the section or team leader.	____	____
3 Protected yourself against physical and other hazards.	____	____
4 Isolated hostile civilian(s), as required.	____	____
5 Employed no more force than was necessary to control the situation.	____	____
6 Established and maintained control of the situation.	____	____

Evaluation Guidance: Score the Soldier GO if all performance measures are passed. Score the Soldier NO-GO if any performance measure is failed. If the Soldier scores a NO-GO, show the Soldier what was done wrong and how to do it correctly.

References
Required
Related: ATP 3-22.40, FM 27-10, TC 7-98-1

181-COM-1001

Conduct Operations According to the Law of War

Conditions: You are a Soldier assigned to a deployed unit which has a mission that requires you to be actively involved in operations that are governed by the Law of War. As a Soldier, you are responsible for identifying, understanding, and complying with the provisions of the Law of War, including the Geneva and Hague Conventions and the 10 Soldier's Rules. You are also responsible for identifying necessary actions to prevent Law of War violations from occurring.

Standards: Identify, understand, and comply with the Law of War. Identify problems or situations that violate the policies and take appropriate action, including notifying appropriate authorities, so that expedient action may be taken to correct the problem or situation.

Special Condition: None

Special Standards: None

Special Equipment:

Cue: None

*Note:*None

Performance Steps

1. Identify the key elements of the Law of War.
    a. Describe how the Hague Convention and Geneva Conventions pertain to combat operations.
    b. Describe International Customary Law of War.
2. Describe the responsibilities of U.S. Soldiers to obey the Law of War.
3. Identify the basic principles of the Law of War.
    a. Define Military Necessity.
       (1) Describe a Legitimate Military Target.
    b. Define Unnecessary Suffering.
    c. Define Discrimination and Distinction.
    d. Define Proportionality.
4. Identify the "10 Soldier's Rules".
    a. Soldiers only fight enemy combatants.
    b. Soldiers treat humanely all who surrender or are captured.
    c. Soldiers do not kill or torture detained personnel.
       (1) List the 5 S's and T.
       (2) Describe humane treatment.
       (3) Describe respect and protect.
    d. Soldiers collect and care for the wounded.
    e. Soldiers do not attack protected places or persons.
       (1) Identify protected persons.
       (2) Identify protected places.
    f. Soldiers do not attack medical personnel, facilities or equipment.
    g. Soldiers destroy no more than the mission requires.
    h. Soldiers treat civilians and noncombatants humanely.

      i. Soldiers do not steal. Soldiers respect private property and possessions.

      j. Soldiers should do their best to prevent violations of the Law of War.

      k. Soldiers report all violations of the Law of War to their superior.

   5. Identify actions to prevent Law of War violations.

      a. List actions to protect civilians/noncombatants.

      b. List actions to protect civilians/noncombatants.

      c. List actions to protect prisoners of war, retained persons and detainees.

      d. List actions to protect medical transports and facilities.

      e. List actions to prevent engagement of unlawful targets.

      f. List actions to prevent excessive use of force.

      g. List actions to prevent the unauthorized use of medical service symbols, flag of truce, national emblems, and enemy insignia/uniforms.

      h. List actions to prevent unnecessary destruction and seizure of property.

      i. List actions to prevent unnecessary suffering and harm.

      j. List actions to enforce the rights and responsibilities of EPWs, and detainees.

Evaluation Preparation:

Setup: Evaluate this task at the end of Law of War training.

Brief Soldier: Tell the Soldier that he or she will be evaluated on his or her ability to identify, understand, and comply with the Law of War, including the Geneva and Hague Conventions and the 10 Soldier's Rules. Tell the Soldier that he or she will also be evaluated on his or her ability to identify problems or situations that violate the Law of War and take appropriate action to prevent Law of War violations do not occur.

Performance Measures	GO	NO GO
1 Identified the key elements of the Law of War.	_____	_____
2 Described the responsibilities of U.S. Soldiers to obey the Law of War.	_____	_____
3 Identified the basic principles of the Law of War.	_____	_____
4 Identified the "10 Soldier's Rules".	_____	_____

Performance Measures	GO	NO GO
5 Identified actions to prevent Law of War violations.	_____	_____

Evaluation Guidance: Score the Soldier GO if all performance measures are passed. Score the Soldier NO GO if any performance measure is failed. If the Soldier scores NO GO, show the Soldier what was done wrong and how to do it correctly.

References:
Required:
Related: AR 27-1, FM 27-10

191-COM-0008

Search an Individual in a Tactical Environment

Conditions: You are assigned the mission of searching an individual for weapons or contraband, given surgical/disposable gloves, a person to provide overwatch, and a translator if available. You have the authorization to search, and the person may or may not have weapons or contraband concealed on his/her person. This task should not be trained in MOPP.

Standards: Search an individual, locating weapons and contraband on the person, while maintain control of the individual throughout the search. Determine the final course of action based on the situation and the result of the search.

Special Condition: Males will search males and females should search females whenever possible. If a female searcher is not available, consider using a doctor, medic or a designated person from the local population to pull clothing tight while you observe.

Cue: None

> **WARNING**
> The searcher must avoid crossing the line of sight or fire of the overwatch during the person search.

References:

> **CAUTION**
>
> Searching a person requires two searchers working together. One searcher conducts the physical search while the other provides overwatch and observes both the searcher and the person being searched. The person providing overwatch should be placed at a 45 degree angle out of the subject's reach.

Performance Steps

1. Determine which type of search to perform based on the situation.
Note: This should be done in conjunction with an interpreter or language handbook of the local population, if available.

   a. Stand-up search with hand restraints.

   b. Stand-up search without hand restraints.

   c. Frisk search.

   d. Prone search.

   e. Strip search. If a strip search is required, it is conducted in a place of confinement/privacy by a qualified person.)

2. Conduct a stand-up search with hand restraints (hand irons or flexicuffs).

   a. Direct the subject to—

     (1) Turn and face away from you.

     (2) Spread his feet until you say "Stop."

     (3) Point his toes outward.

     (4) Bring his hands behind his back with palms out and thumbs upward.

     (5) Stand still.

Note: If the subject resists, attempts to escape, or must be thrown down before the search is complete, restart the search from the beginning.

   b. Ensure that the overwatch Soldier is in the correct position.

   c. Approach the subject cautiously, apply hand restraints, and maintain positive control of the subject throughout the search.

   d. Position yourself behind the subject and remain balanced, with your front foot forward and rear foot to the outside of the subject's feet.

   e. Search the subject's headgear.

     (1) Remove the headgear carefully.

     (2) Bend the seams of the headgear before crushing to detect hidden razor blades or similar items.

     (3) Complete the search of the headgear, and place it on the ground.

   f. Search the subject using the pat-and-crush method in the following sequence:

Note: Mentally divide the body into two parts, and repeat the search for both sides in the same sequence, overlapping areas in the center.

     (1) The head and hair.

     (2) The selected side from arm to shoulder.

     (3) The neck and collar. Bring neckwear worn by the subject to the back, and carefully look for weapons or contraband.

References:

(4) The selected side of the back to the waist.

(5) The selected side of the chest to the waist.

Note: When searching females, check the bra by pulling out the center far enough to allow concealed weapons or contraband to fall out.

(6) The waistband, from the front to the middle of the back.

(a) Bend the material and then crush it to detect razor blades.

(b) Check between the belt and the pants, the pants and the undergarment, and the undergarment and the skin.

(7) The selected side of the buttocks.

Note: Squat when searching the lower half of the subject's body so you are not placed in an unbalanced position.

(8) The selected side hip, abdomen, and crotch.

(9) The selected side leg from the crotch to the top of the shoe.

(10) The selected side shoe. Check the top edge of the boot or shoe by carefully inserting a finger in the top edge to feel for weapons.

g. Reverse the position of your feet, and search the opposite side.

3. Conduct a stand-up search without hand restraints.

Note: The decision to place hand restraints on the subject before searching must be based on the situation and according to the local Standard Operating Procedures (SOP) or policy. It is always safer to have the subject restrained before searching, but in some instances, you may not be authorized. Obtain guidance from your supervisor if you are unsure.

a. Direct the subject to —

(1) Raise his arms above his head, lock his elbows, and spread his fingers with palms facing you.

(2) Turn so that his back is toward you.

(3) Spread his feet apart (more than shoulder width), with his toes pointed out.

(4) Interlock his fingers and place his hands on the crown of his head.

b. Ensure that the overwatch Soldier is in the correct position.

c. Position yourself behind the subject and remain balanced, with your front foot forward and rear foot to the outside of the subject's feet.

d. Search the headgear.

(1) Direct the subject to raise his interlocked hands off his head.

(2) Remove the headgear.

(3) Direct the subject to return his interlocked hands to his head.

(4) Bend the seams of the headgear before crushing to detect hidden razor blades or similar items. Complete the search of the headgear, and place it on the ground.

e. Search the subject using the pat-and-crush method in the following sequence:

(1) Grasp two fingers of the subject's right hand, or both of the interlocked hands with your left hand. Apply pressure, and pull the subject slightly backward to keep the subject off balance.

(2) The head and hair.

(3) The selected side from arm to shoulder.

References:

(4) The neck and collar. Bring neckwear worn by the subject to the back, and carefully look for weapons or contraband.

(5) The selected side of the back to the waist.

(6) The selected side of the chest to the waist.

Note: When searching females, check the bra by pulling out the center far enough to allow concealed weapons or contraband to fall out.

(7) The waistband, from the front to the middle of the back.

(a) Bend the material and then crush it to detect razor blades.

(b) Check between the belt and the pants, the pants and the undergarment, and the undergarment and the skin.

(8) The selected side of the buttocks.

Note: Squat when searching the lower half of the subject's body so you are not placed in an unbalanced position.

(9) The selected side hip, abdomen, and crotch.

(10) The selected side leg from the crotch to the top of the shoe.

(11) The selected side shoe. Check the top edge of the boot or shoe by carefully inserting a finger in the top edge to feel for weapons.

f. Reverse the search to the subject's opposite side.

(1) Grasp two fingers of the subject's left hand with your opposite hand or grasp the interlocked hands with both hands without releasing the subject's right fingers.

(2) Reverse the position of your feet.

(3) Search the opposite side of the subject's body in the same manner as the right side.

4. Conduct a frisk search.

a. Position the subject. Direct the subject to —

(1) Raise his arms above his head, lock his elbows, and spread his fingers with palms facing you.

(2) Turn so that his back is toward you.

(3) Spread his feet apart (more than shoulder width), with his toes pointed out.

(4) Interlock his fingers and place his hands on the crown of his head.

b. Ensure that the overwatch Soldier is in the correct position.

c. Conduct the frisk similar to the stand-up search, except use the massaging method rather than the crushing method to locate weapons.

Note: Although the main intent of a frisk is to ensure that the subject is not carrying a weapon, other contraband found can still be used against the subject. However, be prepared to justify your actions in legal proceedings that result from the search.

(1) Conduct the frisk by searching the outside of the garments only.

(2) Do not search the subject's pockets or waistband unless the pat down suggests the presence of a weapon.

(3) Ask permission to search items that the subject is carrying, such as a purse or backpack. If given permission, search the items in a manner consistent with looking for weapons. If the subject refuses permission to

References:

search the items, or if you are unsure if you are authorized to search these items, contact the military police desk sergeant for further guidance.

(4) Stop the search if a weapon is found, and take whatever measures necessary for your safety before continuing the search.

5. Conduct a prone search.

a. Position the subject.

(1) Command the subject to face you, to raise his arms above his head, lock his elbows, and spread his fingers with the palms facing you.

(2) Visually check the subject's hands for evidence of weapons.

(3) Order the subject to turn around and drop to his knees.

(4) Search the back of the subject's hands for evidence of weapons.

(5) Direct the subject to lie on his stomach, extend his arms straight out to the sides with the palms up, and place his forehead on the ground.

(6) Tell the subject to spread his/her legs as far as possible, turn his/her feet outward, and keep his/her heels in contact with the ground.

Note: Positioning the subject as described is dependent upon the subject following your directions. If the subject refuses, you may have to take his/her to the ground using physical force or whatever alternate use of force is authorized by your PMO.

b. Ensure that the assistant, if available, is in front of and to one side of the subject, opposite the side that is to be searched first.

c. Apply the hand restraints.

(1) Approach the front of the subject at about a 45° angle.

(2) Squat and place your knee that is nearest the subject between his/her shoulder blades.

(3) Direct the subject to put the arm nearest you, behind him/her, with the palm facing up.

(4) If applying handcuffs-

(a) Grasp the subject's hand in a handshake hold and put the first handcuff on it.

(b) Hold the handcuff chain along with the belt or waistband of the subject's trousers, direct the subject to put his other hand behind him, with the palm facing out, and apply the other handcuff.

(5) If applying flexicuffs put the flexicuff around the first hand, holding it along with the waistband of the subject's trousers. Finish by completing the application with the other hand.

d. Search the subject.

Note: Refer to the task special conditions in reference to searching members of the opposite sex.

(1) Hold the center of the hand restraint, and lift the subject's arms slightly. Search the area in the small of the back and any area the subject can reach. Release the chain and stand.

(2) Move to the area of the subject's waist and face the subject's head. Squat, but do not rest your knee on the ground or on the subject. Pivot, if required to conduct the rest of the search.

(3) Remove the subject's headgear.

References:

    (a) Bend the seams before crushing to determine if razor blades or similar devices are hidden.

    (b) Place the headgear on the floor or ground.

    (4) Search the subject's head and hair.

    (5) Search the subject from fingers to shoulders. Search the collar and neck area (pull any neckwear to the subject's back), and remove anything that could be used as a weapon.

    (6) Search the subject's back from shoulder to waist on the side nearest you.

    (7) Grasp the inside of the subject's closest elbow, and pull the subject toward his side, just high enough to search the front without the subject being completely placed on his side. Then, search the front from shoulder to waist. Check the bra area on female subjects.

    (8) Switch hands while controlling the subject's elbow without changing position.

    (9) Search the subject from waist to knee, including the crotch.

    (10) Return the subject to the facedown position, and release the elbow. Remind the subject to keep his feet spread and his heels on the floor.

    (11) Tell the subject to raise his foot by bending at the knee.

    (12) Grasp the subject's foot, and search from the knee up. Check the top of the footwear by inserting a finger in the top edge and feeling for evidence of weapons. You must also check the edges and soles.

    (13) Tell the subject to put his foot back down.

    (14) Stand and move to the subject's unsearched side. Move around the subject's head, but do not walk between the subject and the assistant military police Soldier.

    (15) Ensure that the assistant moves to the side opposite of the side being searched.

    (16) Squat beside the subject, with your body facing the same direction as the subject's head.

    (17) Complete the search of the unsearched side using the same method.

    (18) Help the subject stand once the search is complete by turning him onto his side facing away from you. Have him bring his knees up to his chest. Grasp his arms, and assist the subject to his knees and then to his feet.

    6. Determine the next course of action based on the situation and the result of the search.

    a. If weapons or contraband are found initiate chain of custody documentation, and maintain control of the individual.

    b. If no weapons or contraband are found proceed in accordance with local SOP or as directed by your supervisor.

Evaluation Preparation:

References:
Setup: Provide the Soldier with the items listed in the conditions. Provide a scenario that requires a subject to be searched. Have a role player play the part of the subject. Provide weapons and/or contraband for the role player to conceal on his person. Provide a role player to act as an assistant (not required to test the task).

Brief Soldier: Tell the Soldier to determine the type of search to perform based on the scenario given or directions from the evaluator. Tell the Soldier to perform all of the steps of the search unless otherwise directed. Tell the role players to follow all directions given by the Soldier.

Performance Measures	GO	NO GO
1 Determined which type of search to perform based on the situation.	_____	_____
2 Conducted a stand-up search with hand restrients.	_____	_____
3 Conducted a stand-up search without hand restrients.	_____	_____
4 Conducted a frisk search.	_____	_____
5 Conducted a prone search.	_____	_____
6 Determined the next course of action based on the situation, the result of the search, and/or directions given by the commander or the immediate supervisor.	_____	_____

Environment: Environmental protection is not just the law but the right thing to do. It is a continual process and starts with deliberate planning. Always be alert to ways to protect our environment during training and missions. In doing so, you will contribute to the sustainment of our training resources while protecting people and the environment from harmful effects. Refer to FM 3-34.5 Environmental Considerations and GTA 05-08-002 ENVIRONMENTAL-RELATED RISK

References:
ASSESSMENT. Environmental protection is not just the law but the right thing to do. It is a continual process and starts with deliberate planning. Always be alert to ways to protect our environment during training and missions. In doing so, you will contribute to the sustainment of our training resources while protecting people and the environment from harmful effects. Refer to FM 3-34.5 Environmental Considerations and GTA 05-08-002 ENVIRONMENTAL-RELATED RISK ASSESSMENT

Safety: In a training environment, leaders must perform a risk assessment in accordance with ATP 5-19, Risk Management. Leaders will complete a DD Form 2977 DELIBERATE RISK ASSESSMENT WORKSHEET during the planning and completion of each task and sub-task by assessing mission, enemy, terrain and weather, troops and support available-time available and civil considerations, (METT-TC). Note: During MOPP training, leaders must ensure personnel are monitored for potential heat injury. Local policies and procedures must be followed during times of increased heat category in order to avoid heat related injury. Consider the MOPP work/rest cycles and water replacement guidelines IAW FM 3-11.4, Multiservice Tactics, Techniques, and Procedures for Nuclear, Biological, and Chemical (NBC) Protection, FM 3-11.5, Multiservice Tactics, Techniques, and Procedures for Chemical, Biological, Radiological, and Nuclear Decontamination. Arching a person requires two searchers working together. One searcher conducts the physical search while the other provides overwatch and observes both the searcher and the person being searched.

Evaluation Guidance: Score the Soldier GO if all measures are passed (G). Score the Soldier NO-GO if any measure is failed (F). If the Soldier fails any measure, show him how to do it correctly.

References:
Required: ATP 3-39.12
Related:

159-COM-2026

Identify Combatant and Non-Combatant Personnel and Hybrid Threats

Conditions: In a field, military operations in urban terrain (MOUT), or garrison environment, where a Soldier is required to demonstrate an understanding of the various personnel in an Operational Environment (OE). Standard MOPP 4 conditions do not exist for this task. See the MOPP 4 statement for specific conditions.

Standards: Identify the combatant and non-combatant personnel and hybrid threats within an OE.

Special Condition: None

Safety Risk: Low

MopP 4: N/A

Cue: None

Note: None.

Performance Steps

1. Identify the combatant and/or noncombatant personnel within an OE.
    a. Identify Armed Combatants:
        (1). Regular military forces
        (2). Internal security forces.
        (3). Insurgent organizations.
        (4). Guerilla organizations.
        (5). Private security organizations
        (6). Criminal organizations
    b. Identify Unarmed Combatants:
        (1). Unarmed nonmilitary personnel who may decide to support hostilities-recruiting, financing, intelligence- gathering, providing targeting information, supply brokering, transportation, courier, information warfare (videographers), improvised explosive device (IED) fabricators.
        (2). Unarmed combatants may possibly be affiliated with paramilitary organizations.
        (3). Includes support that takes place off the battlefield.
        (4). Other examples of unarmed combatants-medical teams, media (local, national, international), non- governmental organizations/private voluntary organizations (NGOs/PVOs), Trans-national corporations, foreign government and diplomatic personnel, internally displaced persons (IDPs), transients, local populace.
    c. Identify the following types of Noncombatant.
        (1) Media personnel.
        (2). Humanitarian Relief Organizations.
        (3). Multinational Corporations.
        (4). Criminal organizations.
        (5). Private Security Organizations.
        (6). Other Noncombatants and Civilian Population Support.
        (7). Information Warfare elements.
2. Identify Hybrid Threats within an OE.
(Asterisks indicates a leader performance step.).

Evaluation Preparation:

Setup: Score the soldier GO if all performance measure are passed. Score the soldier NO-GO if any performance measure is failed. If the soldier scores NO-GO, show the soldier what was done wrong and how to do it correctly.

Brief Soldier: This task may be evaluated at the end of OE training as well as during a field training exercise.

Performance Measures	GO	NO GO
1 Identify Combatants.	_____	_____
2 Identify Paramilitary Forces.	_____	_____
3 Identify Insurgents.	_____	_____
4 Identify Terrorists.	_____	_____
5 Identify Drug and Criminal Organizations.	_____	_____
6 Identify Hybrid Threats.	_____	_____
7. Identify Noncombatants	_____	_____

Evaluation Guidance: Score the Soldier GO if all performance measures are passed. Score the Soldier NO-GO if any performance measure is failed. If the Soldier scores a NO-GO, show the Soldier what was done wrong and how to do it correctly.

References ADP 3-0, TC 7-100
Required
Related: ATP 3-22.40, FM 27-10, TC 7-98-1

This page intentionally left blank.

Battle Drills

React to Contact:
071-COM-0502 Move under Direct Fire
071-COM-0030 Engage Targets with an M16 series Rifle/M4 Series Carbine
071-COM-0608 Use Visual Signaling Techniques
113-COM-1022 Perform Voice Communications
071-COM-0503 Move over, Through, or Around Obstacles (Except Minefields)
071-COM-0510 React to Indirect Fire while dismounted
071-COM-0513 Select Hasty Fighting Positions
071-COM-0501 Move as a member of a Fire Team
071-COM-4407 Employ Hand Grenades

Establish Security at the Halt:
071-COM-0801 Challenge Persons Entering your Area
191-COM-5140 Search a vehicle for Explosive Devices or Prohibited items as an Installation Access Control Point
071-COM-0815 Practice Noise and Light Discipline
113-COM-2070 Operate SINCGARS Single-Channel (SC)
113-COM-1022 Perform Voice Communications
171-COM-4079 Send a Situation Report (SITREP)
171-COM-4080 Send a Spot Report (SPOTREP)
071-COM-0513 Select Hasty Fighting Positions
071-COM-0608 Use Visual Signaling Techniques
052-COM-1361 Camouflage Yourself and Individual Equipment
071-COM-4408 Construct an Individual Fighting Position

Perform Tactical Combat Casualty Care:
081-COM-0101 Request Medical Evacuation
081-COM-1003 Perform First Aid to Clear an Object stuck in the Throat of a Conscious Casualty
081-COM-1005 Perform First Aid to Prevent or Control Shock
081-COM-1023 Open an Airway
081-COM-1032 Perform First Aid for Bleeding and/or Severed Extremity
081-COM-1046 Transport a causality
081-COM-1007 Perform First Aid for Burns
113-COM-1022 Perform Voice Communications
191-COM-0008 Search an Individual in a Tactical Environment

React to Ambush:
Near-
052-COM-1271 Identify visual indicators of an IED
052-COM-3261 React to an IED attack
071-COM-0512 React to Hand-to-Hand Combat
071-COM-0030 Engage targets with M4/M16 Rifle

071-COM-4407 Employ hand grenades
071-COM-0501 Move as a member of a team
071-COM-0502 Move under direct fire
071-COM-0513 Select Hasty fighting positions
071-COM-0608 Use visual signal techniques
113-COM-1022 Perform voice communications

Far-

052-COM-1270 React to an IED attack
071-COM-0501 Move as a member of a team
071-COM-0513 Select Hasty fighting positions
113-COM-1022 Perform voice communications
071-COM-0608 Use visual signal techniques
071-COM-0510 React to Indirect Fire dismounted
071-COM-0030 Engage Targets with individual weapon

Appendix B

Proponent School or Agency Codes

The first three digits of the task number identify the proponent school or agency responsible for the task. Record any comments or questions regarding the task summaries contained in this manual on a DA Form 2028 (*Recommended Changes to Publications and Blank Forms*) and send it to the proponent school with an information copy to:

Commander, U.S. Army Training Support Center
ATTN: ATIC-ITSC-CM
Fort Eustis, VA 23604-5166.

Table B-1. Proponent School or Agency Codes	
School Code	**Command**
MSCoE CM **031**	U.S. Army Chemical School Directorate of Training/Training Development 464 MANSCEN Loop, Suite 2617 Fort Leonard Wood, MO 65473-8929
MSCoE EN **052**	Commandant, U.S. Army Engineer School ATTN: ATSE-DT (Individual Training Division) 320 MANSCEN Loop, Suite 370 Fort Leonard Wood, MO 65473
FCoE **061**	Directorate of Training and Doctrine U.S. Army Field Artillery School ATTN: ATSF-D Fort Sill, OK 73503-5000
MCoE **071**	Commandant, U.S. Army Infantry School ATTN: ATSH-OTSS Fort Benning, GA 31905-5593
AHS **081**	Department of Training Support ATTN: MCCS-HTI 1750 Greeley Rd, Ste 135 Fort Sam Houston, TX 78234-5078
SCoE **091 (OMMS)** **093 (OMEMS)**	U.S. Army Combined Arms Support Command (CASCOM) Training Directorate USACASCOM, ATTN: ATCL-TD 2221 Adams Avenue., Suite 2018 Fort Lee, VA 23801-1809

Table B-1. Proponent School or Agency Codes	
School Code	*Command*
SCoE 101	Commander, US Army Quartermaster Center and School ATTN: ATSM-MA Fort Lee, VA 23801-5000
SCoE 113	Commander, USA Signal Center & School ATTN: ATZH-DTM-U Fort Gordon, GA 30905-5074
MCoE 171	Commander, USA Armor Center and School ATTN: ATZK-TDT-TD 204 1ST Cavalry Regiment Road Fort Knox, KY 40121-5123
JAG 181	Commandant, Judge Advocate General Legal Center and School ATTN: JAGS-TDD 600 Massie Road Charlottesville, VA 22903-1781
MSCoE MP 191	Commandant, United States Army Military Police School ATTN: ATSJ-Z 401 MANSCEN Loop, Suite 1068 Fort Leonard Wood, MO 65473-8926
APAC 224	Director, Army Public Affairs Center 6 ACR Road, Bldg 8607 ATTN: SAPA-PA Fort Meade, MD 20755-5650
ICoE 301	Commander, USA Intelligence Center & Fort Huachuca 550 Cibeque Street, Suite 168 ATTN: ATZS-TDS-I Fort Huachuca, AZ 85613-7002
JFK 331	U.S. Army JFK Special Warfare Center and School Fort Bragg, NC 28310-5000

Table B-1. Proponent School or Agency Codes	
School Code	*Command*

SCoE
551
	U.S. Army Combined Arms Support
	  Command (CASCOM) Training
	  Directorate
	USACASCOM, ATTN: ATCL-A
	2221 Adams Avenue
	Fort Lee, VA 23801-2102

Fort Jackson
SRT
	U.S. Army Training Center, Fort Jackson
	  Director of Basic Combat Training
	  (DBCT), Doctrine and Training
	  Development, ATTN: (ATZJ-DTD)
	  4325 Jackson Blvd. Fort Jackson, SC
	  29207-5315

This Page intentionally left blank.

GLOSSARY

Section I
Acronyms & Abbreviations

5-Cs	confirm, clear, call, cordon, and control
AAL	additional authorization list
ACE	air combat element (NATO);analysis and control element;armored combat earthmover;assistant corps engineer;aviation combat element (USMC); Avenger Control Electronics
ADA	air defense artillery; audio distribution amplifier; American Dietetic Association
AO	area of operations
APC	armored personnel carrier; activity processing code
AVPU	alertness, responsiveness to vocal stimuli, responsiveness to painful stimuli, unresponsiveness
BII	basic issue items
BSI	body substance isolation; Base Support Installation
CASEVAC	casualty evacuation
CBRN	chemical, biological, radiological, and nuclear
CPR	cardiopulmonary resuscitation
CSF	cerebrospinal fluid
CWIED	command wire improvised explosive device
DETCORD	detonator cord
EH	explosive hazards
EOF	escalation of force
EPW	enemy prisoner of war
F	frequency; fail; Fahrenheit; full; failed; Feeder; FMC
FM	field manual; frequency modulatedmodulation; flare multiunit; force module
FMC	full mission-capable; field medical card
FMI	field manual-interim; Failure Mode Identifier (indicates type of failure experienced by components. FMI has been adopted from SAE practice of J1587 diagnostics)
FOB	forward operating operationsoperational base; Free on Board
FW	Fixed Wing; Framework
GTA	graphic training aid
HQ	headquarters
IBA	individual ballistic armor
IED	imitative electromagnetic deception;improvised explosive device
JP	joint publication
LACE	liquid, ammunition, casualty, and equipment
LN; ln	local national; lane
MANPADS	man-portable air defense system

MEDEVAC	medical evacuation
METT-TC	A memory aid used in two contexts: (1) In the context of information management, the major subject categories into which relevant information is grouped for military operations: mission, enemy, terrain and weather, troops and support available, time available, civil considerations. (2) In the context of tactics, the major factors considered during mission analysis. [Note: the Marine Corps uses METT-T: mission, enemy, terrain and weather, troops and support available-time available.] (FM 6-0)
MIJI	meaconing, interference, jamming, and intrusion
MOI	message of interest; Material of Interest; memorandum of instruction; mechanism of injury
NBC	nuclear, biological, and chemical
NGO	nongovernmental organization; national government organization
NPA	net pay advice; nasopharyngeal airway
OAKOC	observation and fields of fire, avenues of approach, key terrain, obstacles, and cover and concealment
P	needs practice; pass; passed; barometric pressure; mean radius of curvature; positions; power; Propagated Booster; PMC
PBIED	person-borne improvised explosive device
PIR	priority intelligence requirements; priority information requirements
PZ	pickup zone
RCIED	radio controlled improvised explosive device
ROE	rules of engagement
RPG	rocket-propelled grenade
RTO	radio/telephone operator
RW	rotary wing; readwriter
SALUTE	size, activity, location, unit, time, and equipment
SMCT	Soldier's Manual of Common Tasks
SOI	signal operating/operation instructions
SOP	standing operating procedure
STP	shielded twisted pair; Soldier Training Publication; spanning-tree protocol; Soldier training plan
SURG	surgeon
SVBIED	suicide vehicle-borne improvised explosive device
TC	technical coordinator; training circular; track commander; tank commander; tactical commander; technical configuration
TCCC	tactical combat casualty care
TTP	tactics, techniques, and procedures
US	United States; ultrasound
VBIED	vehicle borne improvised explosive device
VOIED	victim-operated improvised explosive device

cont	continued; continuous; continuous fire; controlled substance
pnt	patient

This Page intentionally left blank.

REFERENCES

Required Publications

Required publications are sources that users must read in order to understand or to comply with this publication.

JOINT PUBLICATIONS
Most joint publications are available online at
www.dtic.mil/doctrine/new_pubs/jointpub.htm
JP 1-02. *Joint Publication 1-02. Department of Defense Dictionary of Military and Associated Terms, 15 June 2015.*
ACP 125 US SUPP-1. *Communications Instructions Radiotelephone Procedures for Use by United States Ground Forces.* 1 October 1985.
ACP 131 *Communications Instructions* Operating Signals, April 2009

ARMY PUBLICATIONS
Army regulations are available on the APD Web Site
(www.apd.army.mil)
ADP 3-0. *Unified Land Operations.* 10 October 2011.
ADP 7-0. *Training Units and Developing Leader.* 23 August 2012.
ADRP 1-02 *Terms and Military Symbols, 2 February 2015*
ATP 3-11.37. *Multi-Service Tactics, Techniques, and Procedures for Chemical, Biological, Radiological, and Nuclear Reconnaissance and Surveillance {MCWP 3-37.4; NTTP 3-11.29; AFTTP 3-2.44}.* 25 March 2013.
ATP 3-37.34. *Survivability Operations, 28 June 2013*
ATP 3-39.12. *Law Enforcement Investigations.* 19 August 2013
ATP 4-02.2. Medical Evacuation. 12 August 2014.
ATP 4-02.285. *Multiservice Tactics, Techniques and Procedures for Treatment of Chemical Agent Casualties and Conventional Military Chemical Injuries. 18 September 2007.*
ATP 4-25.13. *Casualty Evacuation.* 15 February 2013
ATP 5-19. *Risk Management.* 21 August 2014
FM 3-11.3. *MultiService Tactics, Techniques, and Procedures for Chemical, Biological, Radiological, and Nuclear Contamination Avoidance.* 2 February 2006.
FM 3-11.4. *MultiService Tactics, Techniques, and Procedures for Nuclear, Biological, and Chemical (NBC) Protection.* 2 June 2003.

FM 3-11.5. *MultiService Tactics, Techniques, and Procedures for Chemical, Biological, Radiological, and Nuclear Decontamination.* 4 April 2006

FM 3-22.9. *Rifle Marksmanship M16-/M4-Series Weapons.* 12 August 2008.

FM 3-24. *Counterinsurgency.* 13 May 2014.

FM 4-25.11. *First Aid.* 23 December 2002.

FM 6-99. *US Army Report and message Formats.* 19 August 2013.

FM 21-60. *Visual Signal.* 30 September 1987.

TB 9-2320-280-35-2 Installation Instructions for Systems Single Channel Ground and Airborne Radio System (SINCGARS), 17 September 2005.

TC 3-11-55. *Joint-Services Lightweight Integrated-Suit Technology (JSLIST).* 1 July 2001.

TC 3-21.75. *The Warrior Ethos and Soldier Combat Skills.* 13 August 2013.

TC 3-23.30. *Grenades and Pyrotechnic Signals.* 22 November 2013.

TC 7-100. *Hybrid Threat.* 26 November 2010

TM 3-4230-229-10. *Operators Manual for Decontaminating Kit, Skin: M291, (NSN 4230-01-251-8702).* 2 October 1989.

TM 3-4230-235-10. *Operators Manual for Decontaminating Kit, Individual Equipment: M295 (NSN 6850-01-357-8456).* 21 November 2008.

TM 3-4240-300-10-2. *Operator's Manual for Chemical-Biological Mask: Combat Vehicle, M42 (NSN 4240-01-258-0064) Small, (4240-01-258-0065) Medium, (4240-01-258-0066) Large.* 30 August 1988

TM 3-4240-312-12&P. *Operator's and Unit Maintenance Manual for Mask, Chemical-Biological: Aircraft, M43, Type I (NSN 4240-01-208-6966) Small, (4240-01-208-6967) Medium, (4240-01-208-6968) Large, (4240-01-208-6969) Extra Large; Type II (4240-01-265-2677) Small, (4240-01-265-2679) Medium, (4240-01-265-2678) Large, (4240-01-265-2680) Extra Large.* 30 June 1988.

TM 3-4240-342-10. *Operator's Manual for Mask, Chemical-Biological: Field, M4oA1, (NSN 4240-01-370-382-SMALL) (4240-01-370-3822- Medium) (4240-01-370-3823-Large); Mask, Chemical-Biological: Combat Vehicle, M42A2 (4240-01-4100- Small) (4240-01-413-4101-Medium (4240-01-413-4102-Large).* 15 May 2015.

TM 3-4240-346-10. *Operators Manual for Mask, Chemical Biological Mask: Field, M40A1 (NSN 4240-01-370-3821-Small) (4240-01-370-3822-Medium) (4240-01-*

370- 3823-Large); Chemical-Biological Mask: Combat Vehicle, M42A2 (4240-01-4100). 15 May 2015.

TM 3-4240-542-13&P. *Operator and Field Maintenance Manual (including Repair Parts and Special Tools List) for Mask, Chemical-Biological: Joint Service General Purpose, Field, M50 Purpose, Field, M50.* 30 May 2008.

TM 3-6665-311-10. *Operators Manual for Paper Chemical Agent Detector: M9.* 31 August 1998.

TM 3-6665-426-10. *Operator's Manual for Detector Kit, Chemical Agent: M256A2 (NSN: 6665-01-563-7473.* 2 November 2009.

TM 3-9905-001-10, *Operators manual for Marking Set, Contamination: Nuclear, Biological, Chemical (NBC) (NSN 9905-12-124-5955).* 23 August 1982

TM 9-1005-319-10. *Operator's Manual for Rifle, 5.56 MM, M16A2 W/E (1005-01-128-9936); Rifle, 5.56 MM, M16A3 (1005-01-357-5112); Rifle, 5.56 MM, M16A4 (1005-01-383-2872); Carbine, 5.56 MM, M4 W/E (1005-01-231-0973); Carbine, 5.56 MM, M4A1 (1005-01-382-0953).* 30 June 2010.

TM 3-4230-236-10. *Operator's Manual for Decontamination System, Sorbent.* 29 June 2001.

TM 9-1330-200-12. *Operators and organizational Maintenance Manual for Grenades {TM 1330-12/1A}* 17 September 1971.

TM 10-8415-209-10, *Operators Manual for Individual Chemical Protective Clothing.* 31 March 1993.

TM 10-8415-220-10. *Operators Manual for Joint Service Lightweight Integrated Suit technology (JSLIST) Chemical Protective Ensemble.* 28 July 2008.

TM 11-5820-890-10-1. *Operators Manual for SINCGARS Ground Combat NET Radio, ICOM Manpack Radio AN/PRC-119A (NSN 5820-01-267-9482) (EIC: L2Q) Short Range Vehicular Radio AN/VRC-87A (5820-01-267-9480) (EIC: L22) Short Range Vehicular Radio with Single Radio Mount AN/VRC-87C (5820-01-304-2045) (EIC: GDC) Short Range Vehicular Radio with Dismount AN/VRC-88A (5820-01-267-9481) (EIC: L23) Short Range Vehicular Radio with Dismount and Single Radio Mount AN/VRC-88C (5820-01-304-2044) (EIC: GDD) Short Range/Long Range Vehicular Radio AN/VRC-89A (5820-01-267-9479) (EIC: L24) Long Range Vehicular Radio AN/VRC-90A (5820-01-268-5105) (EIC: L25) Short Range/Long Range Vehicular Radio with Dismount AN/VRC-91A (5820-01-267-9478) (EIC: L26) Long Range/Long Range Vehicular Radio*

AN/VRC-92A (5820-01-267-9477) (EIC: L27). 1 September 1991.

TM 11-5820-890-10-3. *Operators Manual for SINCGARS Ground Combat NET Radio, NON-ICOM Manpack Radio, AN/PRC-119 (NSN 5820-01-151-9915) (EIC: L2A) Short Range Vehicular Radio AN/VRC-87 (5820-01-151-9916) (EIC: L2T) Short Range Vehicular Radio (With Single Radio Mount) AN/VRC-87D (5820-01-351-5259) (EIC: TBD) Short Range Vehicular Radio with Dismount AN/VRC-88 (5820-01-151-9917) (EIC: L2U) Short Range Vehicular Radio with Dismount (with Single Radio Mount) AN/VRC-88D (5820-01-352-1694) (EIC: TDB) Short Range/Long Range Vehicular Radio AN/VRC-89 (5820-01-151-9918) (EIC: L2V) Long Range Vehicular Radio AN/VRC-90 (5820-01-151-9919) (EIC: L2W) Short Range/Long Range Vehicular Radio with Dismount AN/VRC-91 (5820-01-151-9920) (EIC: L2X) Long Range/Long Range Vehicular Radio AN/VRC-92 (5820-01-151-9921) (EIC: L2Y).* 1 September 1992.

TM 11-5820-890-10-8. *Operator's Manual for SINCGARS Ground Combat Net Radio, ICOM Manpack Radio, AN/PRC-119A (NSN 5820-01-267-9482) (EIC: L2Q), Short Range Vehicular Radio AN/VRC-87A (5820-01-267-9480) (EIC: L22), Short Range Vehicular Radio with Single Radio Mount AN/VRC-87C (5820-01-304-2045) (EIC: GDC), Short Range Vehicular Radio with Dismount AN/VRC-88A (5820-01-267-9481) (EIC: L23), Short Range/Long Range Vehicular Radio AN/VRC-89A (5820-01-267-9479) (EIC: L24), Long Range Vehicular Radio AN/VRC-90A (5820-01-268-5105) (EIC: L25), Short Range/Long Range Vehicular Radio With Dismount AN/VRC-91A (5820-01-267-9478) (EIC: L26), Short Range/Long Range Vehicular Radio AN/VRC-92A (5820-01-267-9477) (EIC: L27) Used with Automated Net Control Device (ANCD) (AN/CYZ-10) Precision Lightweight GPS Receiver (PLGR) (AN/PSN-11) Secure Telephone Unit (Stu) Frequency Hopping Multiplexer (FHMUX).* 1 December 1998.

TRADOC PAM 600-4. *The Soldier's Blue Book.* 2014.

RELATED PUBLICATIONS

Related publications are sources of additional information. They are not required in order to understand this publication.

ARMY PUBLICATIONS
Army regulations are available on the APD Web Site

(www.apd.army.mil)

AR 27-1. *Legal Services, Judge Advocate Legal Services (*RAR 001, 09/13/2011).* 30 September 1996.

ATP 3-22.40. *NLW Multi-Service Tactics, Techniques, and Procedures for the Tactical Employment of Nonlethal Weapons. 13 February* 2015.

DA Pamphet 750-8. *The Army Maintenance Management System (TAMMS)User's Manual.* 22 August 2005.

GTA 05-08-002. *Environmental-Related Risk Assessment. 31* October 2013.

FM 2-91.6. *Soldier Surveillance and Reconnaissance: Fundamentals of Tactical Information Collection.* 10 October 2007.

FM 3-34.5. *Environmental Considerations.* 16 February 2010.

FM 6-02.53. *Tactical Radio Operations.* 5 August 2009.

FM 22-6. *Guard Duty.* 17 September 1971.

FM 27-10. *The Law of Land Warfare.* 18 July 1956.

TC 3-25.26. *Map Reading and Land Navigation.* 15 November 2013.

TC 3-25.150. *Combatives.* 24 September 2012

TC 3-34.489. *The Soldier and the Environment.* 8 May 2001

TC 7-98-1. *Stability and Support Operations Training Support Package.* 5 June 1997.

Recommended Readings
Operational Law Handbook
UNIT SOI Unit/Unit's Signal Operation Instructions (SOI)
Geneva and Hague Convention, *Laws of War.*
  http://www.loc.gov/rr/frd/Military_Law/pdf/ASubjScd-27-1_1975.pdf

Referenced Forms
Department of the Army Forms
DA Forms are available on the APD Web Site (www.apd.army.mil)

DA Form 1594. *Daily Staff Journal or Duty Officer's Log.*

DA Form 2028. *Recommended Changes to Publications and Blank Forms.*

DA Form 2404. *Equipment Inspection and Maintenance Worksheet.*

DA Form 5164-R. *Hands-on Evaluation.*

DA Form 5165-R. *Field Expedient Squad Book.*

DA Form 5517-R. *Standard Range card*

DA Form 5988-E. *Equipment Inspection Maintenance Worksheet (EGA).* Printed forms are available through normal supply channels.

DA Form 7656. *Tactical Combat Casualty Care (TCCC) Card.*

Department of Defense Form

DD Forms are available on the APD Web Site
(www.dtic.mil/whs/directives/infomgt/forms/index.htm)

DD Form 1380. *US Field Medical Card.*

DD Form 2977. *Deliberate Risk Assessment Worksheet.*

By order of the Secretary of the Army:

RAYMOND T. ODIERNO
General, United States Army
Chief of Staff

Official:

GERALD B. O'KEEFE
Administrative Assistant to the
Secretary of the Army
1521901

DISTRIBUTION:

Active Army, Army National Guard, and U.S. Army Reserve: To be distributed in accordance with the initial distribution number (IDN) 111447, requirements for STP 21-1-SMCT.

WARRIOR ETHOS

The Warrior Ethos forms the foundation for the American Soldier's spirit and total commitment to victory, in peace and war, always exemplifying the ethical behavior and Army Values. Soldiers put the mission first, refuse to accept defeat, never quit, and never leave behind a fellow American. Their absolute faith in themselves and their comrades makes the United States Army invariably persuasive in peace and invincible in war.

35040173R00187

Made in the USA
Middletown, DE
16 September 2016